T0294853

A EUROPEAN JUST TRANSITION FOR A BETTER WORLD

The Green European Foundation (GEF) is a European-level political foundation whose mission is to contribute to a lively European sphere of debate and to foster greater citizen involvement in European politics. The GEF strives to mainstream discussions about European policies, both within and beyond the Green political family. The foundation acts as a laboratory for new ideas, contributing to cross-border political education and offering a platform for cooperation and exchange at the European level.

About the Just Transition project

This book is part of the Green European Foundation's transnational Just Transition project. The project investigates the issues around transforming an economy from being extractive to regenerative in a just and equitable way, in the hope of finding the necessary support among the population. The project is focused on collecting and sharing insights on the development of future-proof politics and policies that are developed in a sensitive way, keeping in mind local specificities. The project is coordinated by Oikos (Belgium) on behalf of the GEF. The project partners are Green House Think Tank (UK), the Institute for Political Ecology (IPE, Croatia), Sunrise (North Macedonia), Transición Verde (Spain), the Federation of Young European Greens (FYEG), Visio (Finland) and Polekol (Serbia). In 2020 and 2021 these organizations carried out various activities in their home countries, including events, publications and interviews, to increase awareness of the importance of a just transition. This book is the outcome of that collaboration, and it showcases the main challenges and opportunities around just transition.

Editor: Dirk Holemans (Oikos)

Authors: Natalie Bennett (Green Party peer), Joaquín Nieto (ILO Spain), Dirk Holemans (Oikos), Elina Volodchenko (Oikos), Lyda Fernanda Forero, Daniel Chavez, Anya Namaqua Links, Rand El Zein, Raúl Gómez (Transición Verde), Vedran Horvat (IPE), Dragan Djunda (Polekol), Aleksandar Gjorgjievski (Sunrise), Simo Raittila (Visio), Sara Matthieu (Member of the European Parliament), Sean Currie (Scottish Young Greens), Anne Chapman (Green House), Robert Magowan, Adrián Tóth (Green European Foundation).

Project Coordinator: Adrián Tóth, Green European Foundation

Cover design and layout: Miriam Hempel

Copies of the book can be ordered by sending an email request to info@gef.eu

---

Contact the Green European Foundation:
Social seat address: Rue du Fossé 3, L-1536 Luxembourg
Brussels office: Mundo Madou, Avenue des Arts 7-8, 1210 Brussels, Belgium
Tel: +32 2 329 00 50
Email: info@gef.eu
Website: www.gef.eu

A European Just Transition for a Better World

Edited by
Dirk Holemans

GEF GREEN EUROPEAN FOUNDATION

Copyright © 2022 by Green European Foundation

Published by London Publishing Partnership
www.londonpublishingpartnership.co.uk

Published in association with the
Green European Foundation www.gef.eu

European Political Foundation — N° BE 089.337.8512

Published with the financial support of the European Parliament
to the Green European Foundation. The views expressed in this
book are solely those of the authors and do not necessarily reflect
the views of the European Parliament.

ISBN: 978-1-913019-58-7 (hbk)
ISBN: 978-1-913019-59-4 (ePDF)
ISBN: 978-1-913019-60-0 (ePUB)

A catalogue record for this book is available
from the British Library

Typeset in Adobe Garamond Pro by
T&T Productions Ltd, London
www.tandtproductions.com

Contents

Foreword

Natalie Bennett

To be a Briton introducing a collection of essays about a just, green transition across Europe and around the world feels fitting. Fitting because Prime Minister Margaret Thatcher's unjust, ideologically driven transition of the UK economy away from coal, cutting an economic and social swathe through the heart of what are now some of the comparatively poorest areas of Europe, continues to serve as a cautionary tale for Polish and German mining communities, Italian steel cities and French farmers.

Yet, as this volume makes clear, change does not have to be negative. There is no shortage of case studies of the ways in which transitioning our economy to a sustainable future can clean up our air and waters and can also give us healthier food, fitter bodies and more satisfying and secure lives. Thatcher set out to destroy an opposing political force that was grounded in coal communities. The fate of those communities was not collateral damage so much as intentional destruction.

To borrow a favourite phrase of the Iron Lady, 'there is no alternative' but to transition away from our current economic model – and fast. The climate emergency, existential as its threat is, represents just one way in which we are exceeding the limits of our single fragile planet. The one thing that is certain in the coming years is massive, foundational change, because the position we are in today is profoundly unstable – economically, socially and environmentally. And that is actually good news.

I often ask audiences to conduct a thought experiment. Imagine that we had smoothly functioning, fair, decent societies in which everyone had a steady, reliable income sufficient to meet their needs; a warm, comfortable, genuinely affordable home; high levels of physical and mental health; tasty, healthy and accessible food;

flourishing natural wildlife and clean air and water; and a secure international environment. If in that world it was discovered that we faced a climate emergency and a nature crisis and that we needed to radically change our way of living, that would be a *really* difficult political challenge.

Instead we are in a world that has to change, quickly, for environmental reasons, but also for social justice reasons. When nearly a billion people are regularly going to bed hungry, when the young are struggling to find secure employment and the old are battling poverty, when the state of nature is parlous, the oceans are turning into a plastic soup and the climate is discernibly heading for disaster, we should be seeing people leaping at positive change – for a just transition.

That they are demonstrably not doing so, even in societies with genuine democratic choices, is not a measure of any kind of innate human conservatism or fear of change. It is not the result of any kind of brilliance among the Far Right forces that seek to win support and votes through painting this as a difficult, dangerous world in which we have to grab scarce resources for 'us' and 'ours', whoever 'us' is defined to be, while pushing away the 'others', all while pandering to rich individuals and multinational companies with tax cuts and deregulation.

It is rather a failure of our collective politics to paint an attractive, believable, comprehensible picture of what a just, green society looks like – that it is not just 'business as usual with different technology'. We need to demonstrate that reducing working hours as an alternative to acquiring more and more 'stuff' is a huge gain. To show that clean air and water, verdant wildlife and cities that do not groan with the sound of internal combustion engines are healthier and happier. To show that genuine security – freedom from fear of want – can be delivered and that it is indeed the natural next step after decades of progressive recognition of human rights.

That is what the following pages do: they offer hope of a better life for all, collectively lived within the limits of this one fragile planet, while nature is restored and temperatures are kept below 1.5 °C above pre-industrial levels.

Please enjoy and explore these pages, and do not keep them to yourself. For the most important message of all is that the future is in all of our hands. History is not pre-written, but made. And to make a good future possible, far more people need to be engaged and enthused, so that communities and individuals make politics what they do, not what is done to them.

Introduction

Dirk Holemans

As I am making the final changes to this introduction, the media are reporting that on New Year's Day 2022 world records for anomalous temperatures will be broken all over the world. Whether they live in Belgium, Great Britain or Portugal, Europeans have experienced the warmest end of year ever recorded, and on other continents the temperatures have been unprecedented too. In South America, Paraguay measured a record of 45.6 °C. The entirety of 2021 was a disaster year in terms of extreme weather phenomena, as a result of destructive human interference with the biosphere. And although climate disruption and nature destruction is felt everywhere, the effects are unevenly distributed. The population of Africa in particular pays the price, even though it is responsible for barely 4% of greenhouse gas emissions. Every year, all over the world, more people and more communities see their jobs, livelihoods and habitats threatened or destroyed by prolonged drought, heatwaves and extreme rainfall. All these observations can be summarized in one powerful sentence that Greta Thunberg used to describe the situation: 'Our house is on fire.'

That means we have to act with unprecedented speed to stop further global warming and the further collapse of biodiversity. It must happen in this decade. But that is not the end of the challenge. Not only do we have to move quickly along the required transition path, but we must also travel far along it, for merely greening the existing economic system will not get us anywhere. The authoritative reports of both the Intergovernmental Panel on Climate Change (IPCC) and the Intergovernmental Science-Policy Platform on Biodiversity and Ecosystem Services (IPBES) are unanimous on this point: only transformative changes, in all parts of our society, can provide an adequate response. Radical change in the short term is what is needed if we want to leave our children a liveable world and alleviate the suffering that the current human-induced changes in the biosphere are already causing. Furthermore, this challenging transition has to be realized in a world that is also characterized by growing inequality, rapid digitalization and, of course, the Covid-19 crisis.

We can learn two important things from the past: when things change substantially in society, there are always winners and losers, and change always provokes resistance because of the uncertainty it creates. These are two basic facts that must be taken into account if we are to increase the chances of making the necessary transformational changes. The two things are also connected.

It is clear that the regions and groups in society that are already in a precarious situation are the most vulnerable in a process of transition towards a world that learns to respect the planetary boundaries. Just think of the case of an underprivileged person renting a poorly insulated house and driving an old car because there is no good public transport to take them to work. If, in an injudicious manner and in the name of an ambitious climate policy, a carbon tax is then introduced, or if transport is brought under the EU's Emissions Trading System without also introducing mitigating social policies, then there is a chance that the already low quality of life of people in poverty will be further reduced.

As a consequence, there may be an understandable uprising of people who feel they have been treated unjustly, as was the case with the 'yellow jacket' protests in France. These citizens did not turn against environmental measures but against their president, who had first abolished a wealth tax for the rich and then introduced a tax on fuel under the guise of ecological concerns. This reminds us of the core of political ecology: the search for a good life for all, within the limits of the planet. It is about offering everyone the opportunities to flourish in their community and the right to the necessary basic services that allow them to make their own choices in life, to feel safe in a changing world and to work together for the emancipation of all. This is the core of what a just transition means: the perspective from which to develop new policies and politics, public institutions, industries and commons to make the necessary transformative changes possible.

This brings us to the second thing that we can learn from history: the risks that arise when societal groups feel insecure, left alone and fearful of the future. To give a concrete example again, well-meaning measures in climate policies such as subsidizing the purchase of electric cars can be counterproductive: they give money to people who can probably already afford to buy such an expensive car, while people in the lowest-income groups simply cannot afford to buy even a cheap car. So if there is not, at the same time, sufficient investment in public transport, existing inequality can even be exacerbated by climate policies. History also shows us that the uncertainty created by ill-conceived policies and measures is, unfortunately, a perfect breeding ground for authoritarian leaders and political parties (parties that most of the time support fossil fuel industries and reject just transition politics); this is something that we also see happening every day.

In this sense, a just transition is not simply the wishful thinking of certain groups in society; it is a meeting of necessity and ethics. It is necessary to shape the transition in such a way that all groups feel included, because otherwise the chances are high that the necessary support will never materialize and that resistance will only grow. It is our moral duty to leave no one behind and to ensure that the opportunities for everyone to lead a good life increase – with, of course, priority support for groups and regions that are already struggling and being exploited, or that we know will be badly affected, such as regions with mining or fossil fuel industries.

As we argue in this book, a just transition requires nothing more or less than a new social contract. In the twentieth century, the social contract of the welfare state in Europe immensely improved the quality of life of large parts of the population, but it was based on predatory capitalism that plundered the rest of the world and its nature. A new contract for Europe – this time a social–ecological contract – can only be written from a global perspective, taking into account human rights and respect for nature worldwide. At the same time, the need for this major transformation requires a great deal of customization. The differences between European regions are enormous, not least in terms of income disparity between and within countries. So, too, are the financial, innovative and structural capacities to shape and implement the transformative changes that are needed. This can be clearly seen in the different contributions of this book. Only far-reaching forms of solidarity and cooperation can provide an answer to this.

It is from these perspectives that we critically appreciate the European Commission's Green Deal. Compared with the policy of the previous Commission, it is a true break from the earlier neoliberal perspective. Climate ambitions have been greatly increased, and unprecedented funds have been provided. Now it is a question of getting the money to the right targets, and not just to the sectors and companies that are best at lobbying. This requires that transition policies are also justice policies from the outset and that there is room for bottom-up participation and substantial investment in a social dialogue, where civil society, including trade unions and environmental groups, is represented. A just transition is both about creating new jobs and about increasing social justice, as well as being about providing new answers to complex problems.

As we explain in this book, the concept of 'just transition' has already been part of climate discussions for a long time: it features in the preamble of the Paris Agreement, for example. Meanwhile, we are already several United Nations (UN) Climate Change Conferences further along, and at the end of 2021 the 26th UN Climate Change Conference of the Parties (COP26) took place in Glasgow. The good news is that just transition was present in the debates and texts linked to COP26. Also, it was not only trade unions but also indigenous peoples and environmental movements that put the issue on the agenda. But attention in itself is not enough. Without denying that progress was made at COP26 by taking just transition more seriously, much more still needs to be done. Signing pledges is one thing – implementing them in a way that respects redistributional, recognitional and procedural justice is something different. In concrete terms, when the hard decisions are made, they respond to the following questions: who gets a seat at the table, how are decisions made in a democratic way, and how are the burdens and benefits shared?

This relates to the deep need for a new social–ecological contract that we mentioned earlier. In essence, just transition is not about the illusion that greening our capitalist economy and making our society a little bit less unequal would provide an answer to climate change. We are facing multiple crises that can only be dealt with by transformative – and thus systemic – changes. It is, in the end, about the following fundamental questions: how do we do things justly and what is a just society within planetary boundaries?

PART I

Setting the Stage

Just transition in the climate agenda: from origins to practical implementation

Joaquín Nieto*

GENESIS

The adoption of the concept of 'just transition' by the United Nations in its sustainable development policies and in the framework of the climate agenda has been a long process. As far as the latter is concerned, the official adoption of the concept did not take place until 2015, in the Paris Agreement, at the 21st Conference of the Parties (COP21), the preamble of which contains the notion of just transition. The agreement recognizes that countries may be affected not only by climate change *per se,* 'but also by the impacts of the measures taken in response to it', emphasizing the importance of protecting people and their jobs in the process of change, and of putting in place concrete measures to mitigate the effects on the sectors and territories that are most vulnerable. The agreement adds that it is up to each country to articulate just transition processes in the context of its own energy transition.

It was not until 23 years after the international community adopted the United Nations Framework Convention on Climate

*Joaquín Nieto was Director of the International Labour Organization (ILO) Office for Spain (2011–2021) and the first president of the Sustainlabour Foundation (2004–2008). This chapter is largely a revision and update of work published in the monographic issue of *Papeles de Economía Española* 163 (2020) on the transition to a low-carbon economy, which was prepared by the author together with Ana Belén Sánchez (regional specialist on green jobs for the ILO Regional Office for Latin America and the Caribbean) and Julieta Lobato (research professor at the University of Buenos Aires).

Change (UNFCCC), in 1992, that the concept of just transition was incorporated into the international climate agenda. It has been a long process in which both the maturation of the social dimension of climate change in multilateral organizations and, above all, trade union initiatives have been key.

In the multilateral framework, the social and labour dimension in relation to environmental issues appeared as early as 1972, at the UN Conference on the Human Environment in Stockholm, and then again later at the UN Conference on Environment and Development in Rio de Janeiro, in 1992, and at the World Summit on Sustainable Development in Johannesburg, known as Rio+10, in 2002. But it was not until 2012 that the United Nations adopted the concept of just transition for the first time in a resolution approved at Rio+20: 'The Future We Want'.* At this international conference, again held in Rio, there was an express will to merge the social and environmental agendas under the perspective of sustainable development. In addition to renewing their commitment to sustainable development, the heads of state and high representatives of the countries that make up the United Nations system that were present recognized 'the importance of a just transition, including programmes to help workers to adjust to changing labour market conditions' and the need for job creation opportunities and decent work for all, committing to 'work towards safe and decent working conditions and access to social protection and education'.

This recognition of just transition was not a coincidence but the result of intense trade union activity during the preparation process for Rio+20 and at the conference itself. Activity was coordinated by the International Trade Union Confederation (ITUC),† under the leadership of Anabella Rosemberg, and by the Sustainlabour Foundation,‡ directed by Laura Martín Murillo. The two organizations

---

* See https://bit.ly/3JKIqL4.

† The ITUC, created in 2006 by the merger of the International Confederation of Free Trade Unions (ICFTU) and the WCL, has 180 million members, belonging to more than 300 trade union organizations in more than 150 countries.

‡ Sustainlabour, the International Labour Foundation on Sustainable Development, was founded in 2004 to support the environmental activity of trade unions

had been working together for almost a decade with two primary goals. Firstly, they promote the climate and environmental agendas in the trade union and labour spheres, at both the national and international levels, as well as in multilateral negotiations, supporting the processes of incorporating trade unions into the sustainable development agenda. And secondly, they advocate for the incorporation of the social and labour dimensions with a just transition approach into the climate and environmental agendas at both national and international levels.

The ILO, as the UN's specialized agency on the subject, plays a special role in this process of increasing the understanding of the relationship between the environment and labour and of the interactions between the climate agenda and the labour agenda in multilateral organizations.

The 1992 Rio summit had an impact on all institutions, including the ILO. In 1994 one of the main architects of the Rio summit, UN Under-Secretary-General Maurice Strong, approached the ILO on the occasion of its 75th anniversary to ask it to integrate environmental notions into its role in favour of social justice and development. The request fell on fertile ground, as the ILO had been incorporating environmental references since the 1970s, mainly in relation to the working environment in terms of air pollution and the use of hazardous substances, but also in the preservation of the environment, which was recognized as being of fundamental value to indigenous peoples in Convention 169 (1989).*

The integration of environmental notions has continued to progress and be consolidated, notably in the areas of green jobs and climate change, with an early adoption of a just transition approach.

In 2008 the ILO, the UN Environment Programme (UNEP), the ITUC and the International Organization of Employers produced a report titled 'Green jobs: towards decent work in a sustainable,

---

at national and international level. Its first president was Joaquín Nieto, the leader of the Spanish trade union Comisiones Obreras at the time, and the last was Sharan Burrow, General Secretary of the International Trade Union Confederation.

*ILO Convention 169, from 1989, on indigenous and tribal peoples. Available at https://bit.ly/3bgGlYg.

low-carbon world' (International Labour Organization 2008b). The report represented the first global study on the impacts of the green economy on the world of work, and it was the first publication to unite two traditionally separate worlds that were often seen as enemies under the false premise that environmental protection measures necessarily resulted in job destruction. The report brought together the world of work (trade unions, employers and the ILO) with the world of the environment (through UNEP) to jointly identify the employment impacts of a more environmentally sustainable development model.

The report argued that just transition is essential to decarbonize the economy and to move towards sustainable and egalitarian societies. Furthermore, it underlined the fact that the most radical and profound changes needed to achieve sustainable economies are related to redefining most jobs. The report also highlighted that: 'Climate change itself, adaptation to it and efforts to arrest it by reducing emissions have far-reaching implications for economic and social development, for production and consumption patterns and thus for employment, incomes and poverty reduction.'

The report also pointed out that 'green jobs'* are not automatically decent jobs. This means that many jobs that reduce pressure on the environment do not necessarily have the characteristics of decent work, i.e. work with rights and without discrimination, in safe and healthy conditions, with sufficient pay to live in dignity and with social protection. Along these lines, the ILO governing body stated the need to govern the energy transition process, in order to take advantage of the opportunities presented by the changes and to not succumb to the challenges (International Labour Organization 2008a). The ILO also drew attention to the secondary nature of the socio-labour implications of the climate agenda, because 'employment and development benefits are essential to make mitigation measures technically feasible, economically viable, socially acceptable and politically sustainable'.

---

* The report states that green jobs 'reduce the environmental impact of enterprises and economic sectors, ultimately to levels that are sustainable'.

In 2013 the International Labour Conference adopted the 'Resolution concerning sustainable development, decent work and green jobs'.* In 2015, in the process of implementing that resolution, the ILO developed and adopted jointly and tripartite (i.e. with the formal and full participation of governments, trade union organizations and employers' organizations from around the world) the 'Policy guidelines for a just transition to environmentally sustainable economies and societies for all' (International Labour Organization 2015). These guidelines constituted the global roadmap for implementing just transition.

The energy and environmental transition had a major presence in the debates promoted by the ILO on the future of work on the occasion of its centenary (International Labour Organization 2017c). Firstly, this transition has been identified as one of the megatrends that needs to be considered because – together with others such as the technological revolution, demographic change and the incorporation of the gender dimension in any social agenda – it has been reshaping the world of work. Moreover, the tripartite Centenary Declaration committed to directing its efforts to ensure 'a just transition to a future of work that contributes to sustainable development in its economic, social and environmental dimensions' (International Labour Organization 2019a). This action must be carried out through policies that place people, their needs and their rights at their core, given that the impact of climate change affects low-income countries and the most vulnerable populations to a greater extent, and particularly those who work in the so-called informal economy, which translates into a lack of access to health services, employment, pensions and protection against accidents. Young people, women, the rural population and migrants are the most affected groups.

---

* Available at https://bit.ly/3hjE0PS.

THE CONFLICTIVE EVOLUTION OF THE TRADE UNION MOVEMENT AS AN ARCHITECT OF JUST TRANSITION

One of the first public mentions of the just transition concept was made in 1993 by American trade unionist Tony Mazzocchi when he called for opportunities and financial assistance for workers who were displaced from their jobs by the implementation of environmental protection policies. At the same time, Canadian trade unions were using the concept in their fight against the asbestos industry.

Regarding the climate agenda, the concept is used as a way to overcome the potential adverse effects of greenhouse gas mitigation measures (such as moving away from coal and other fossil fuels) on employment. Neither the UNFCCC nor its first operational instrument, the Kyoto Protocol, contemplated this dimension at the time. The Convention rightly contemplated the principle of 'common but differentiated responsibilities' and considered the 'respective capabilities and their social and economic conditions' to address climate change and to adopt mitigation measures 'with a view to minimizing adverse effects on the economy, on public health and on the quality of the environment, of projects or measures undertaken by them to mitigate or adapt to climate change'.* However, the UNFCCC never names direct policies on employment or social justice. As a result, no channels were created to incorporate the employment issue into the climate negotiations. The search for such channels, incorporating just transition into the Convention's development instruments, became a trade union priority in its participation as a civil society observer party in the negotiations. But the road to adopting this strategy was far from easy.

At COP3 in 1997, where the Kyoto Protocol was adopted, workers' representatives expressed opposing positions: on the one hand, North American trade unionists, which contained a large contingent of mining representatives, opposed the adoption of mitigation targets; and on the other hand, the European Trade Union Confederation supported the adoption of the Protocol and proposed to

---

* See https://unfccc.int/resource/docs/convkp/conveng.pdf.

fully develop the measures in favour of renewable energies and other measures provided for in Article 2 of the Kyoto Protocol. This was intended to overcome, through the creation of new jobs, the adverse effects on employment of the measures to close mines and coal-fired power plants – measures that were necessary to meet the Protocol's mitigation targets for industrialized countries. But the process of convergence between European and North American trade unions in relation to Kyoto was very complex and long-winded.* It required the development of social dialogue processes between the social partners in each country, in order to anticipate the positive and negative effects of mitigation measures.

The trade union agenda advanced considerably at the beginning of the twenty-first century with the increasingly coordinated participation of a growing number of national trade union centres in successive climate COPs coordinated by the ICFTU (the forerunner to the ITUC) and with the establishment of Sustainlabour, promoted by the ICFTU as an instrument of support for trade union organizations on climate and environmental issues. In this context, the trade union proposal gained ground: the only way to ensure that the climate agenda moved forward and that the ecological transition process was successful was a fair labour process. Thus, for example, at COP13 in Bali in 2007, the trade union delegation, made up of more than 80 representatives from 22 countries, presented a proposal to position the employment issue within the framework of the negotiations in the following terms:

> Employment transitions should be studied and anticipated so as to guarantee social justice. Accompanying measures (including the promotion of decent and green jobs and social protection systems)

---

* While European trade unions supported the Kyoto Protocol from the beginning and negotiated its implementation, the US unions' position against the Kyoto Protocol continued for years. In conversations between European and North American unions in New York, in which I participated back in 2004, US trade unionists told us that for them the Kyoto Protocol was death, and that the just transition represented nothing more than a coffin for a "nice burial". But this position began to change shortly after the flooding of New Orleans in the wake of Hurricane Katrina in August 2005.

need to be designed along with mitigation and adaptation measures. Trade unions propose to place employment, income and pro-poor measures at the centre of the discussions. Employment impacts should be incorporated as a variable in all scenarios.*

At the same time, alliances began to be forged with various European governments and other parts of the world, notably with representatives from Argentina, making proposals to formally incorporate just transition into the text of the new instruments to be adopted under the Convention. This led to the Paris Agreement at COP21 in 2015, which finally incorporated this concept.

PERCEPTIONS OF CLIMATE CHANGE IMPACTS ON THE ECONOMY AND EMPLOYMENT

Understanding of the impact of climate change on the economy and employment has increased over time. Various studies have contributed to this change, and it has enabled the economic and social dimensions of the climate agenda to be understood and incorporated into decisions.

The publication of *The Economics of Climate Change: The Stern Review* in 2006 represented a turning point in the acceptance of the economic and social dimensions of the climate agenda, and it established the conditions for a consensus space to emerge for the trade union movement to intervene in favour of just transition. The Stern report warned of the profound negative economic and social consequences of not acting quickly to curb climate change, stating that it could cost up to 20% of GDP per year indefinitely. This was in contrast to investing to avoid it, which would cost as little as 1% of GDP and would have positive economic consequences.

As the Fourth Assessment Report (AR4) from the Intergovernmental Panel on Climate Change in 2007 concluded:

> The benefits and costs of climate change for industry, settlement and society will vary widely by location and scale... In the aggregate,

---

*Extracts from the Trade Union Statement at COP 13, December 2007.

however, net effects are more likely to be strongly negative under larger or more rapid warming.

In this context the European trade union movement also made its contribution through a report titled 'Climate change and employment. Impact on employment in the European Union-25 of climate change and CO_2 emission reduction measures by 2030' (European Trade Union Confederation 2006), which was jointly produced by the Trade Union Confederation's Trade Union Institute for Labour, Environment and Health (Spain), Syndex (France) and the Wuppertal Institute (Germany). Using various measurement methodologies, the report took stock of the impacts of climate change on the different European regions, based on the projection of a moderate change scenario. The European Trade Union Confederation endorsed the report's conclusions about the impact on jobs.

At the same time, Sustainlabour published its 'Guide on climate change', in which it drew attention to the different effects of the environmental transition on employment in various productive sectors, such as agriculture, forestry and ecosystems; on health and human settlements; and on society.

Likewise, the aforementioned report on 'Green jobs: towards decent work in a sustainable, low-carbon world' (International Labour Organization 2018a) represented an extraordinary advance in the understanding both of the interactions between climate change, the economy and employment and of the need to incorporate this issue into the climate agenda.

ILO studies report that each year the increased frequency and intensity of human-related natural disasters decreases the productivity of ecosystems, on which 1.2 billion jobs depend – 40% of global employment. Between 2000 and 2015, natural disasters that were caused or aggravated by humankind caused an annual global loss of 23 million life years – equivalent to 0.8% of a year's work. In turn, the effects of what is known as 'heat stress' result in an annual loss of working hours equivalent to some 80 million jobs (International Labour Organization 2019b).

The ILO has also studied the impact on employment of the implementation of the mitigation commitments made by individual countries under the Paris Agreement. The study estimated that six million jobs will be lost by 2030, but twenty-four million will be created in return. This implies the creation of four new jobs for every job lost. Even more jobs will be created if countries promote a circular economy (International Labour Organization 2018b).

But the situation is much more complex because, although four times as many jobs will be created as are lost, the jobs that are created will not be created in exactly the same place and at exactly the same time as those that will be destroyed. The ILO report shows that out of the 163 economic sectors analysed, most will benefit from net job creation. Among them, the electricity sector stands out, where the switch to renewable energy is estimated to create 2.5 million jobs by 2030, which puts into perspective the 400,000 jobs that will be lost in fossil-fuel-based electricity generation. In addition, only 14 sectors will lose more than 10,000 jobs, and only two sectors (oil extraction and refining) will lose a million jobs or more. In regional terms, net job creation is estimated at 3 million for the Americas, 14 million for the Asia-Pacific region and 2 million for Europe. The region that will be most negatively affected is the Middle East, where the net job loss will be 0.5% due to the importance of oil in the region.

Moreover, the circular economy is a promising sector, as it is estimated to create 6 million additional jobs in the coming years.

On the other hand, environmental degradation particularly affects the most vulnerable groups in society: women, the poor, migrant workers and indigenous peoples, among others. Women, who account for 48% of global labour market participation, occupy only 32% of all jobs in the growing renewable energy sector. Of these, almost 50% are in administrative jobs, while only 28% are in decision-making and science-trained positions. It is obvious that these jobs also have better pay and more favourable working conditions. This gap is also evident in the sustainable construction sector, in the recycling industry, in organic agriculture and in the electric vehicle industry (Sanchez 2019). The situation is exacerbated by the greater difficulties women face in adapting to climate change, in terms of access to financial resources, agricultural resources, land, technology and training (Baruah 2016; Intergovernmental Panel on Climate Change 2014; International Labour Organization

2009; International Labour Organization 2017a; von Hagen and Willems 2012).

It is therefore necessary to conduct the processes of energy transition and ecological reconversion on the basis of decent work and social justice. Thus, it is essential to set the roadmap that this process should follow in a fair way, which implies moving towards an environmentally sustainable economy through the correct and efficient management of the transition, in order to contribute to the achievement of decent work for all, social inclusion and poverty eradication.

There is no doubt that the climate agenda has long been recognised within the international development agenda. This prominence has been boosted by the adoption of the Sustainable Development Goals (SDGs) as part of the UN's 2030 Agenda. Climate change is explicitly addressed in SDG 13 ('Climate action'), whose targets propose mainstreaming climate change action into national policies, strategies and plans, and also improving education, awareness and human and institutional capacity for climate change mitigation.

Many of the other goals and targets are also related to the climate agenda – not only those directly linked to environmental goals, such as water, energy, forests and oceans, but also those related to poverty, health, gender equality, industry, consumption, cities, the economy and employment. The 17 SDGs constitute an agenda of social transformation for environmentally sustainable human development, and the 169 targets that they contain constitute a specific transition strategy for each of the subjects, with all the goals and targets being interrelated.

Specifically with regard to the aspects most closely related to just transition, SDG 8 ('Promote sustained, inclusive and sustainable economic growth, full and productive employment and decent work for all') states that, to achieve sustainable economic development, societies are called upon to create the necessary conditions for people to have access to decent work with quality jobs, stimulating the economy without damaging the environment. It complements this principle in SDG 1 ('End poverty') by categorically stating that economic growth must be inclusive, in order to create sustainable jobs and promote equality, and above all by calling for the extension of basic levels of social protection to all countries in the world.

JUST TRANSITION AS AN INSPIRATION FOR THE 'INTERNATIONAL CLIMATE ACTION FOR JOBS' INITIATIVE

The 2015 Paris Agreement's incorporation of the notion of just transition opened up a process for its adoption by many countries. This momentum was maintained all the way through to the New York climate conference of September 2019, when more than fifty countries signed a declaration supporting the just transition.*

The incorporation of just transition in the Paris Agreement was welcomed by the ILO, whose director general, Guy Ryder, stressed that it was the first time that the need to respect the rights of all people, including workers, in climate transition processes was noted, guaranteeing decent work for all.

Following the Paris Agreement, the UN, with the support of the ILO, has been leading on what is known as 'the social dimension of climate change'.

The first major breakthrough in this area after COP21 in Paris took place at COP24 in Katowice in 2018, where many countries signed the Solidarity and Just Transition Silesia Declaration.†

Subsequently, at the New York climate conference in September 2019, UN Secretary-General António Guterres launched the International Climate Action for Jobs (AC4J) initiative,‡ mandating the ILO to develop this initiative in coordination with social partners, with the aim of encouraging countries to adopt national just transition plans as part of their climate policies.

The initiative's main objective was to put people at the centre of climate action. This means ensuring that climate action is accompanied by the creation of decent work and green jobs, with specific social protection measures for vulnerable groups. It also means providing economic incentives to support private sector technological and energy retrofitting for low-carbon production, and also social dialogue mechanisms to guide the broad agreements that need to

---

* See https://bit.ly/3qGeYOQ.
† See https://bit.ly/2uYLW33.
‡ See www.climateaction4jobs.org/.

be reached along the way, including calls on countries joining the initiative to commit to adopting national just transition plans as part of their climate policies.

The AC4J initiative, initially launched with the participation of 46 countries, was officially presented at COP25 in Madrid.*

Just transition has also been advancing, in a cross-cutting manner, in climate negotiations. The COP25 outcome document reaffirmed, in Decision 1/CP.25, the mandate of the Paris Agreement to take into account the imperatives of a just reconversion or transition of the workforce and the creation of decent work and quality jobs. For example, in the work plan on response measures to address climate change, just transition was included in the following activities: identifying national strategies and good practices for the implementation of low-emission policies; promoting parties' capacities to analyse the impacts of the implementation of response policies; using guidelines and frameworks to assist parties; and exchanging regional, national and sectoral experiences.

On gender, just transition was incorporated into the principles of the new Gender Action Plan (GAP), under the Lima Work Programme on Gender. The reformulation of the GAP was one of the breakthroughs of COP25 due to the disproportionate impact of the climate emergency on women. Incorporating just transition into this plan implies recognizing the existing labour gaps in the labour markets that must be combatted through targeted policies, at the risk of reproducing the same gender inequalities that are currently found in the new sustainable economic designs.

Throughout 2020 and 2021, in the midst of the Covid-19 pandemic, the AC4J initiative – far from coming to a standstill – has continued to be deployed with the aim of accelerating the energy and ecological transition processes around the world. An example of this is the launch of the AC4J initiative in Africa – something that took place last April. In 2020 the initiative's International Advisory Council was formed, with the participation of ministers from Argentina, Costa Rica, Spain, France, Ghana, Indonesia, Samoa and Senegal. The Council is chaired by Samoa's Minister of Labour and

---

* The event registration can be accessed at https://bit.ly/3uEtqYO.

Spain's Vice-President and Minister for Ecological Transition. This co-presidency was a declaration of intent: it brought together (as is usual in the UN) a developing country and an industrialized one, but also a ministry of labour and a ministry of ecology, a territory particularly vulnerable to climate change (such as a Pacific island) and a country that is a pioneer in the application of just transition.

In this post-Covid-19 scenario, the economic reorientation of the EU, which has decided to launch a Green Deal within a framework of greater political ambition to tackle climate change, is encouraging news, with targets of 55% emission reduction by 2030 and zero emissions by 2050, and with multi-billion-euro investments, some of which will go to finance just transition. This is especially welcome, since this approach has been integrated into the European and national plans for economic and social recovery to address the destructive health, economic and employment impacts of Covid-19. The EU's Next Generation* programme, which envisages a combined investment of €806.9 billion, is more than a recovery plan. The incorporation of climate responsibility, energy and ecological transition, social inclusion, and just transition is good news, and it could represent a before and after towards sustainability for the EU and for each of its members.

CONCLUSIONS

This chapter, which focuses on the need to address the employment dimension of climate policies through a just transition, has shown the concept's long journey before its adoption into the official agenda of the framework of the UNFCCC. It has also outlined the role of the political, institutional and, above all, trade union actors who have contributed to its incorporation. Today, just transition is supported by a wide range of sectors, including business (see, for example, We Mean Business Coalition 2019).

The conceptual content of the notion of just transition and the instruments needed to achieve it have also been described: notably the ILO guidelines on the subject and the European policies and

---

* Available at https://ec.europa.eu/info/strategy/recovery-plan-europe_en.

investments geared towards economic and social recovery, inspired by the Green Deal.

To facilitate the processes of ecological and energy transition with sufficient scope and speed to avoid catastrophic climate change, political decisions and agreements are essential, but they are not the only agreements needed. To overcome the social obstacles to change, a broad social agreement is necessary through a framework of social dialogue and citizen participation that includes the social dimension and guarantees a just transition. Only in this way will the energy and ecological transition be possible and the necessary change be effective. Those economies that are better placed in this transition will have better opportunities; those that take longer will have more difficulties and greater negative social impacts. It is for all of us to avoid the socio-environmental collapse of our civilization, which is still a real threat, and to move towards environmentally sustainable and socially just societies.

BIBLIOGRAPHY

Baruah, B. 2016. Renewable inequity? Women's employment in clean energy in industrialized, emerging and developing economies. *Natural Resource Forum* 41(1), 18–29.

Bureau for Workers' Activities. 2018. Just transition towards environmentally sustainable economies and societies for all. Policy Brief, International Labour Organization, Geneva (https://bit.ly/3LFPGJs).

European Commission. 2019. Clean energy for all Europeans. Report, European Commission, Brussels.

European Trade Union Confederation. 2006. Climate change and employment: impact on employment in the European Union-25 of climate change and CO2 emission reduction measures by 2030. Report, ETUC, Brussels (https://bit.ly/3sQcQUS).

European Trade Union Confederation. 2011. ETUC resolution Rio+20: strengthening the social dimension of sustainable development. Report, 19–20 October, ETUC, Brussels (https://bit.ly/34MbK4w).

European Trade Union Confederation. 2017. Resolution on the follow-up of the Paris Agreement on climate change. Report, 27 October, ETUC, Brussels (https://bit.ly/3rWfcSV).

Intergovernmental Panel on Climate Change. 2007. Fourth assessment report (AR4) of the United Nations Intergovernmental Panel on Climate Change (www.ipcc.ch/assessment-report/ar4/).

Intergovernmental Panel on Climate Change. 2014. *Climate Change 2014: Impacts, Adaptation, and Vulnerability.* Cambridge University Press.

International Labour Organization. 2008a. Employment and labour market implications of climate change. Conference Proceedings, 303rd Session, International Labour Office, Geneva (https://bit.ly/3sADK4j).

International Labour Organization. 2008b. Green jobs: decent work in a sustainable, low-carbon world. Report, United Nations Environment Programme (https://bit.ly/353VYBU).

International Labour Organization. 2009. Improving the climate for gender equality too! Website, ILO, Geneva (https://bit.ly/357sumR).

International Labour Organization. 2015. Guidelines for a just transition to environmentally sustainable economies and societies for all. Policy Guidelines, ILO, Geneva (https://bit.ly/3HXi9YV).

International Labour Organization. 2017a. World employment and social outlook: trends for women 2017. Report, ILO, Geneva (https://bit.ly/3rWV4QJ).

International Labour Organization. 2017b. Gender, labour and a just transition towards environmentally sustainable economies and societies for all. Report, ILO, Geneva.

International Labour Organization. 2017c. Work in a changing climate: the green initiative. Report, ILO, Geneva (https://bit.ly/3GWU0AF).

International Labour Organization. 2018a. Green jobs: towards decent work in a sustainable, low-carbon world. Report, ILO, Geneva (https://bit.ly/3N99p57).

International Labour Organization. 2018b. Greening with jobs. World employment and social outlook 2018. Report, ILO, Geneva (https://bit.ly/3oVxVwh).

International Labour Organization. 2019a. ILO century declaration for the future of work. Conference Proceedings, International Labour Organization Conference, Geneva (https://bit.ly/3sQWyem).

International Labour Organization. 2019b. Working on a warmer planet: the impact of heat stress on labour productivity and decent work. Report, ILO, Geneva (https://bit.ly/30vxsTP)

International Trade Union Confederation. 2010. Resolution on combating climate change through sustainable development and just transition. Resolution, ITUC (https://bit.ly/3sRS8nT).

International Trade Union Confederation. 2017. Just transition – where are we now and what's next? ITUC Climate Justice Frontline Briefing (https://bit.ly/3I3wxz1).

Nieto Sainz, J. 2009. *Los desafíos del cambio climático: manual para comprender la agenda del clima.* Madrid: Ediciones GPS (In Spanish).

Nieto Sainz, J. 2018. Notas sobre la Transición Justa. *Cuadernos de Energía* 57, 101–108 (In Spanish).

Organización Internacional del Trabajo and Conama. 2018. La aplicación de las directrices de la OIT sobre transición justa en el contexto de la transición energética española. Report, OIT/Conama (www.conama.org/informeOIT) (In Spanish).

Rosemerg, A. 2010. Building a just transition: the linkages between climate change and employment. *International Journal of Labour Research* 2(2), 125–161.

Ryder, G. 2015. El acuerdo de París: creación de empleo y política climática para una transición justa. Blog Post, 26 October, *El País* (https://bit.ly/3oftbRK) (In Spanish).

Sanchez, A. B. 2019. El empleo verde para las mujeres. Blog Post, 8 March, *AgendaPublica* (https://bit.ly/3oULVGm) (In Spanish).

Stern, N. 2006. *The Economics of Climate Change: The Stern Review.* Cambridge University Press.

Strong, M. 1994. Environment and sustainable development. In *Visions of the Future of Social Justice: Essays on the Occasion of the ILO's 75th Anniversary.* Geneva: International Labour Organization.

United Nations. 1992. United Nations framework convention on climate change. Policy Document (https://unfccc.int/resource/docs/convkp/conveng.pdf).

Von Hagen, M. and Willems, J. 2012. Women's participation in green growth: a potential fully realised? Study, Deutsche Gesellschaft für Internationale Zusammenarbeit.

We Mean Business Coalition. 2019. Climate and the just transition. Blog Post, 18 January (https://bit.ly/3BsMcVY).

Different perspectives on a just transition: from decent jobs in a greener society to a good life for all within planetary boundaries

Dirk Holemans and Elina Volodchenko

INTRODUCTION

Sustainable development – a concept that has dominated public debate on socio-ecological challenges for decades – promised the simultaneous flourishing of people, planet and profit. And yet in recent years, inequality has risen to an all-time high (Piketty 2013). The financial crisis of 2008 provoked the most severe economic shock since the 1930s, pushing millions of people worldwide into poverty (Alexander 2010). In particular, the gap has widened between the incomes of the top 1% and the rest of the population, and the gap between richer regions and the periphery has therefore grown in some countries. Despite the observed economic growth, inequalities have been increasing, due to rising unemployment, underemployment or precarious work (such as forms of part-time work and short-term contracts: see Organisation for Economic Co-operation and Development (2019)). Also, more and more people who have a 'regular' job are no longer able to meet the demands of rising work tempo and productivity, causing burnout and other mental health issues.

Moreover, we have already transgressed four of the nine planetary boundaries – extinction rate (one of two indicators for biosphere integrity), deforestation, atmospheric carbon dioxide (an indicator for climate change) and the flow of nitrogen and phosphorus – thereby exceeding the carrying capacity of the planet in four of the nine critical parameters (Steffen *et al.* 2015). This will affect not only the economic sphere, but also the social, political and societal spheres

(Bollon *et al.* 2020). Food security, employment, working conditions and overall political stability will not be unaffected by global warming. The negative impacts of climate change are expected to affect low-income countries most. However, the impact of such a drastic crisis will not be limited to the Global South; there will be implications for high-income countries in the Global North as well (Koch 2018). Examples of such direct implications include heatwaves, forest fires and rising sea levels. Indirect effects on Europe could include a degraded coastal infrastructure that will hinder transport and shipping. Overall, climate change will disrupt food supply, leading to rising and volatile prices, which in turn could lead to disturbances in global economic networks and chains. This could also lead to rising numbers of (climate) refugees.

Another effect of transgressing planetary boundaries is the rise of epidemics and pandemics (Armelagos and Harper 2009; Koch 2018). Nature is increasingly commodified into raw materials through deforestation and intensive agriculture (Tsing 2008, 2011, 2015). By invading ecosystems, we create more contact points between humans and zoonotic diseases (Vidal 2020). The more we disrupt nature, the more we create habitats in which viruses can be easily transferred to humans. And it is not only Covid-19 that is a consequence of this: Ebola, HIV and dengue are too. The Covid-19 crisis has thrown the need for deep, transformative change into even starker relief. The pandemic and its consequences reinforced inequality worldwide, with people in precarious situations suffering most. It underlined the major shortcomings of hyperglobalization, such as the dramatic effects that our dependency on global supply chains can have on basic rights such as healthcare and food provision (van de Pas 2020). As a global society, we are thus increasingly confronted with multiple crises, requiring us to undertake profound changes (Brand 2016).

In summary, our dominant economic model simply does not function anymore. GDP growth – still the main policy goal of most governments – is now associated with rising inequality, growing feelings of insecurity and loss of well-being, as well as structural degradation (Organisation for Economic Co-operation and Development 2019). Furthermore, decades of neoliberal policies – the so-called

Washington Consensus that prioritizes liberalized markets and criticizes government intervention – created precarity that has led to serious political consequences. Societies have become more fragmented because the trust in established institutions has fallen to an all-time low. Industrial heartlands have no faith in the direction of travel of industrial change, and workers lack trust in their governments (Powell *et al.* 2019). Discontent with politicians and the political system has led to people turning to more extreme alternatives. Populist parties have gained ground in many countries because of this (Holemans 2021).

As our dominant economic system no longer works and its connected welfare state has been severely eroded, it is clear that we need a new vision for the future. Moving from a welfare state to an alternative eco-welfare or well-being society is necessary if we are to prevent our societies becoming more fragmented and in order to rebuild trust in institutions, where we strive to provide a good life for all within planetary boundaries (Koch 2018). If we fail to provide democratic answers, polarizing and authoritarian systems will gain the upper hand (Holemans 2021). The current situation requires a global response, although high-income countries have more responsibility than low-income ones. As we already said, low-income countries in the Global South suffer the most from the negative impacts of climate change while being least responsible for causing it, which creates a double injustice (Walker 2012). In other words, the groups most likely to be affected are those least responsible for exceeding our planetary boundaries. From this perspective, decolonizing the existing extractive economy, in which the Global North relies on cheap resources and labour from the Global South, is a necessity.

Moreover, if rich households within high-income countries do not pay as much as experts deem necessary to carry the financial burden of new climate policies, a triple injustice may even occur (Koch 2018). Low-income households are least able to carry these external costs; for instance, they will feel the impact of rising energy prices the most.

The emergence of the yellow vests (*'gilets jaunes'*) movement in France has also shown good reasons why new environmental policies are only accepted within a framework of just policies. The French president did the exact opposite of this: shortly after abolishing a capital tax on the rich, he introduced an extra levy on fuels for cars, and he did this in a country where, especially in the countryside, workers often have to drive many kilometres in their cars because public transport has been cut. This serves as a strong lesson:

> At the heart of [sustainability narratives and people's reluctance towards them] lies the fear that addressing the monumental climate challenge will inevitably require us to choose between either protecting the planet or protecting workers and the economies that sustain people. The idea that environmental protection and employment protection are incompatible ... cuts across social, political and geographic divides.
>
> Just Transition Research Collaborative (2018, p. 3)

Only a vision of transition that considers the complexity, context dependency and interconnection of the multiple challenges that we face will be able to bring about the radical change that is so urgently needed. This shows that you cannot implement environmental policies without a social framework. Or, in other words: either climate policy will be social or it will not 'be' at all.

The overall goal is striving for a good life for all within planetary boundaries, which requires embedding new production and consumption patterns while remaining aware of possible negative social effects (Holemans 2021). In recent years there has been an increasing focus on the concept of *just transition* to achieve this, as part of a growing trend where concepts such as 'transition' and 'transformation' – terms that are often used interchangeably – are discussed more frequently, reflecting the growing realization that structural changes are needed in our societal architecture. The concept of just transition has been introduced as an overarching framework to guide our transformation into ecological societies in a socially just and equitable way. As this concept develops, labour unions and climate movements are bringing the need for systemic transformation to the fore. Just transition entails fundamental changes, not only to key production and consumption systems such as energy, transport, agriculture and food but also to infrastructure, societal values and politics (Heyen *et al.* 2020). Moreover, it emphasizes the need for a global shift towards a human and fair economic system, with healthy ecosystems, health care, public services, education and culture at its heart.

The core scientific reports of our time also highlight this need. The 2019 report from the Intergovernmental Science-Policy Platform on Biodiversity and Ecosystem Services (IPBES) declares that 'goals for 2030 and beyond may only be achieved through transformative changes across economic, social, political and technological factors'. In the report, transformative changes are defined as 'a fundamental, system-wide reorganization across technological, economic and social factors, including paradigms, goals and values'. For its part, the 2018 Intergovernmental Panel on Climate Change (IPCC)report 'Global warming of 1.5 °C' states the following:

> Limiting warming to 1.5 °C above pre-industrial levels would require transformative systemic change, integrated with sustainable development. Such change would require the upscaling and acceleration of the implementation of far-reaching, multilevel and cross-sectoral climate mitigation and addressing barriers. Such systemic change would need to be linked to complementary adaptation actions, including transformational adaptation, especially for pathways that temporarily overshoot 1.5 °C.
>
> Intergovernmental Panel on Climate Change (2018, p. 315)

An important systemic change is situated at the level of societal paradigms: the redefining of what economic prosperity and well-being really signifies. The indicator we use to measure progress in society today – GDP growth – is no longer adequate (Organisation for Economic Co-operation and Development 2019). GDP growth is no longer correlated with improvements in human well-being. Income is still important, but only to a certain level. Other factors play a role in well-being as well, such as the feeling of security, social networks, relationships, the quality of public goods (such as the health care and education systems) and general trust in society (Hickel 2020). We therefore need other measures for economic and social progress. This will require more than just marginal changes: a total transformation of our modern economic structures is needed. Only with new measures and objectives can new systemic changes be achieved.

The Organisation for Economic Co-operation and Development (OECD) proposes four objectives for policymaking that will lead towards a system that meets both the planet's needs and those of the people. Societies need to work towards each of these four objectives, rather than having only a narrow focus that prioritizes economic growth (Organisation for Economic Co-operation and Development 2019).

The first objective focuses on environmental sustainability and cherishing biodiversity. This is needed to achieve a healthy level of ecosystem services, instead of exploiting natural resources.

The second objective concentrates on increasing well-being. This can be understood in terms of both individual life satisfaction and the improvement of the quality of society as a whole. This therefore means that private consumption cannot be prioritized over common goods, and that differences between groups must always be taken into account.

Well-being is also what the third objective is about: reducing inequality. The ultimate goal is to reduce the wealth gap between certain groups in society. This would be achieved through a reduction in poverty rates, but also through relative improvements in the well-being, incomes and opportunities of the groups that are systematically being marginalized (such as women, members of ethnic minorities and disabled people).

The last OECD objective is system resilience. This has to be implemented to make sure that the new socio-economic model can withstand shocks to the system (in other words, experience no lasting negative impacts), be they financial, environmental or other.

The Covid-19 crisis showed that governments are able to steer their economies when they deem it necessary to do so, and that citizens are willing and able to help those in need. Therefore, a path towards a just transition does not seem unreachable. Using the colossal national and European relief budgets to simply return to business as usual would be a catastrophic mistake. And yet, in many countries, this type of 'recovery' is already taking place.

Things can, however, be done differently. The pandemic offers the opportunity to engineer a comprehensive plan, combining solutions to both the pandemic itself and the imminent threat of global warming and ecological collapse. The Covid-19 crisis has only amplified the need for market regulation, for solid government and for civil society participation. Economic help from governments to businesses suffering losses due to Covid-19 needs to be coupled with binding environmental regulation and legislation.

Against this backdrop, it is clear that we need to find ways to make sure that just transition becomes a guiding principle while we develop pathways and policies for the future. On a European level, further elaboration of the Green Deal's Just Transition Mechanism (JTM), among other things, is necessary but not sufficient. Every part of the Green Deal has to start from the principle of a just transition. On a global level, the importance of just transition is of an even greater magnitude. With soaring poverty and inequality rates worldwide, just transition will mean nothing without a fundamental reassessment of the global economic rationale and of unjustifiable neoliberal and neocolonial regulations (Hickel 2020). It is crucial that we debate and plan for a more profound transition:

> [We need a transition] that could transform the economic and political structures that reproduce and exacerbate inequalities and power asymmetries. Such a radical transition requires a redefinition of economic prosperity and social well-being. At its heart will be the creation of employment that promotes labour rights and improves working conditions while also encompassing gender and racial equality, democratic participation and social justice.
>
> Just Transition Research Collaborative (2018, p. 4).

We must have discussions about the need to decommodify certain societal sectors such as education and care, to create the space for new economic institutions, such as commons, and to promote the role of public services.

THE HISTORICAL ROOTS OF THE DIFFERENT PERSPECTIVES

As mentioned in the previous chapter, the concept of just transition has a broad historical background and, therefore, a broad definition, revealing the different aspects that need to be included in the transformative process, as well as the different implications of the transition on workers, localities, industries and nations (Powell *et al.* 2019). Accordingly, just transition can be interpreted in a very specific way, but also in a very general one. These different perspectives on the concept and how it should best be implemented were shaped historically as well. The origins of just transition lie in reconciling social concerns (of workers and communities) with new environmental policies.

Initially, the goal was particularly focused on improving workers' health and livelihoods while simultaneously taking the environment into account (Just Transition Research Collaborative 2018). Activists and unionists acknowledged at the time that industries were causing climate destruction and health problems. People started to focus on the intersections between health, human rights and environmental conditions, which led to protests against the unjust distribution of environmental hazards (Labor Network for Sustainability n.d.; Cahill *et al.* 2020). Initiatives against the threats of industrial capitalism were organized through the mobilization of workers to secure decent jobs and livelihoods while simultaneously addressing climate change.

The concept of just transition was first mentioned in the 1970s, when workers in the US oil, chemical and nuclear industries were at risk of losing their jobs because of environmental legislation (Galgóczi 2018). At the same time, there was growing discourse about rising jobs versus the climate, fuelled by the neoconservative right (Just Transition Research Collaborative 2018). This discourse shaped the idea that increasing environmental regulations would lead to job losses. As a reaction to this rising discourse, the union movement developed a programme called the 'Superfund for workers' (Mazzocchi 1993). This was a clear signal that policymakers would not have to choose between jobs and the environment in their new measures. In 1993, US trade union leader Tony Mazzocchi advocated for the fund, saying that it would 'provide financial support and opportunities for higher education for workers displaced by environmental protection policies' (Galgóczi 2018).

By the end of the 1990s, just transition had been incorporated into the vocabulary of numerous trade unions in North America, and the concept had also been picked up by international trade unions (Galgóczi 2018). As a result of this provenance, the concept of just transition was initially understood as something that encompassed support programmes for workers whose jobs were threatened by environmental legislation. This notion was shaped by the long-standing philosophy of labour organizations that social concerns should be an integral part of policymaking. Two key features of this original interpretation still survive to this day. Firstly, just transition is not welfare, but rather a comprehensive plan to provide displaced workers not only with financial compensation and security but also, and just as importantly, with proper relocation and adequate retraining opportunities. Secondly, just transition is more than just switching energy sources: we interpret it now as integrated societal adaptation, economic reorientation, appropriate policymaking and equitable resource redistribution (Galgóczi 2018).

In the first decade of the twenty-first century, just transition was increasingly referred to in an international context (Just Transition Research Collaborative 2018). It got picked up in international climate negotiations and had gained traction in the international policy space. However, it was only later that coordinated efforts were undertaken to mainstream just transition. A key promoter of a just transition pathway was the Spain-based Sustainlabour Foundation, a green think tank. The trade unions were still the main promoters of just transition during the global diffusion process, but through active engagement with climate movements in climate debates, the framing of just transition was enlarged.

The focus shifted to the trade union movement's response to the climate debates, centred around global union federations like the International Trade Union Confederation (ITUC):

[In 2009] the ITUC presented Just Transition as 'a tool the trade union movement shares with the international community, aimed

at smoothing the shift towards a more sustainable society and providing hope for the capacity of a "green economy" to sustain decent
jobs and livelihoods for all'.
  United Nations Research Institute for Social Development (2018)

Despite the move towards a greener economy, the focus still had to
be on sustaining decent jobs and livelihoods for all workers. This was
done by shedding light on the social implications of climate mitigation interventions. According to the ITUC, the goal was ultimately
to strengthen the idea even more that environmental and social policies were not inherently contradictory. Trade unions were therefore
an important social partner for the green movement.

Moreover, the concept of just transition was increasingly being
propagated by sections of the climate justice movement. Through
the involvement of the climate movement, the language of just
transition was reinterpreted (Labor Network for Sustainability n.d.).
They advocated for a system change, rather than what had been the
traditional, governmental ways of dealing with climate change. The
belief was upheld that the 'enduring power structures of sovereignty,
capitalism, scientism, patriarchy and even modernity generate and
perpetuate the environmental crisis while consolidating structural
inequalities between the global North and South' (p. 12). The movement was increasingly shaped by a focus on the unequal distributional consequences on different geographical regions.

Many climate justice activists tried to get developed countries to
realize that they were for the most part historically responsible for
climate change, so they would have to act on it by increasing their
efforts towards climate mitigation and by giving developing countries
financial and technological assistance. For example, the commercial-
ized exploitation of resources destroyed ecosystems and livelihoods
through 'elite capture' (United Nations Research Institute for Social
Development 2021). This is the reason why mineral-resource countries
are frequently going through distributional conflicts. A post-colonial
view also reckons with these extraction conflicts, taking indigenous
voices into account. The importance of having a decolonization lens
on the just transition project has therefore increased, widening the
focus beyond the western welfare capitalist system.

As globalization increased between the years 2010 and 2018, just transition entered into mainstream climate change debates (Just Transition Research Collaborative 2018). The concept was adopted by other groups, such as non-governmental organizations (NGOs), UN groups and feminist movements. The term therefore grew in popularity, and the increase in the number of actors involved and in the scope widened interpretations and perspectives of the concept beyond purely labour issues. The importance of an intersectional focus expanded the aim of just transition to include active campaigns seeking justice for different social groups in society by addressing cultural, gendered and racial injustices. The notion remained concentrated around decent jobs and livelihoods for all, but with a more nuanced look at who the 'all' were exactly (with questions such as, 'How will gender, race and disability be considered in an employment setting within a just transition?'). The wider narrative was complemented by a more bottom-up view of change.

In line with this view, the United Nations Research Institute for Social Development (UNRISD) proposed a new social contract for the twenty-first century that takes account of both planetary boundaries and human rights across a broader spectrum (United Nations Research Institute for Social Development 2021). It focuses on intersectional policies that bring in groups that have previously been marginalized, such as women, informal workers, racial and ethnic minorities, LGBTQIA+ persons and migrants.

The new *eco-social contract* focuses on seven topics in particular, through which the UNRISD has tried to reflect the imbalanced relationships that exist in our current model: for instance, between state and citizen, between genders, between the Global North and South, and between humans and the environment and its biodiversity. The goal of a new just transition project should be to redefine and rebalance these hegemonic relationships, inorder to reduce current inequalities, with the key focus being inclusion and sustainability.

A short summary of the seven points

1. There must be a focus on human rights for everyone, including often-marginalized groups.

2. A progressive fiscal contract must be provided in this new framework. It should raise sufficient funds for climate action, and it should distribute the financial burden fairly.
3. The contract should be based on the idea that we transform our economies and societies to promote, for instance, social inclusion, climate mitigation and adaptation.
4. It should also be a contract with nature, taking into account that humans are part of a global ecosystem. We should thus protect ecological processes and biodiversity.
5. A new contract should be decolonized. This means that we should acknowledge historical injustices. A new social contract should also encompass different just transitions and situated knowledge informed by indigenous communities. There should be emphasis on different 'just transitions' instead of just one dominant, homogenous just transition process.
6. Gender justice must be incorporated, stressing equality between men and women. Activities such as production and reproduction should be shared equally by all genders. Furthermore, all sexual orientations and gender identities should be granted equal respect and rights.
7. Participatory, bottom-up approaches are ultimately required to bring about these transformative changes. Alliances between different social groups in society would inform discussions about the changes needed. To stimulate a more inclusive process, a just transition project is required at all levels of society.

So, while just transition originated in the US labor movement, the concept eventually entered the international negotiation space with a wide range of stakeholders (Just Transition Research Collaborative 2018; Cahill *et al.* 2020). This international diffusion brought with it a diversification of perspectives on how to define just transition, incorporating a variety of worldviews, meanings and strategies. The common idea is that we need a sustainable pathway to transition away from an economy powered by fossil fuels, while also considering the impacts of this transition on the jobs and livelihoods of the affected communities. Job losses are likely to occur in regions that are mostly dependent on fossil fuels and where opportunities for economic diversification are limited. That is why a regional focus is very important. This path to a low-carbon economy envisions a proactive role for trade unions.

In time, this perspective was broadened from job creation in a green economy to a radical critique of capitalism (Just Transition Research Collaborative 2018). It focused not only on sustainable pathways but also on ways that could override other systemic forms of oppression, such as racism, sexism, imperialism and islamophobia. It is thus important to see just transition as a process, not just a defined outcome (Powell *et al.* 2019). The perspectives on just transition have broadened the notion over the years, so it may change even more with time. The concept is still primarily used in high-income countries in the Global North, but that is also changing, and the just transition concept is finding more resonance in low-income countries in the Global South. Implementing a just transition framework thus requires constant re-evaluation, through conversations with various social partners and stakeholders.

A JUST TRANSITION FRAMEWORK

> Politicians have a particular responsibility when it comes to Just Transition and climate action. If you take the European Green Deal as an example, this is an intentional state intervention into the market. That then puts the moral responsibility on politicians to manage the social impacts of their political choices on climate and energy policy. The management of these social impacts is essentially the Just Transition framework. But around that agreement there is obviously a role for other organizations. You have the regional governments, the regional and local training providers, educational establishments, society at large and citizens; whether that's through direct citizen engagement or through associations like NGOs. To have strength in the changes you want to realize, you need to engage the people who are affected by the transition.
>
> Judith Kirton-Darling, interview on 2 July, 2021

The Paris Agreement focused on three main transformations to advance environmental sustainability: 'jobs will be transformed, jobs will be lost and jobs will be created' (Union to Union 2020, p. 5). The most important aim of a just transition pathway is to manage the phasing out of high-emitting sectors. This has to be done through moving certain jobs to more sustainable ones that take planetary boundaries into account. Just transition involves implementing measures to help workers who may lose their jobs while also making sure that the new green jobs are good jobs. Moreover, the term indicates that societies need to be more resilient to climate change impacts. In other words, just transition takes the four objectives formulated by the OECD into account. However, the framework does not have specific rules, being a relatively new concept in international policies. Nevertheless, a set of components has to be included to realize a just transition: social protection, organizational health and safety policies, sectorial and active labour market policies, and finally skills and development policies.

In the definition adopted by the ITUC, social dialogue and secure, decent jobs are key:

> A Just Transition secures the future and livelihoods of workers and their communities in the transition to a low-carbon economy. It is based on social dialogue between workers and their unions, employers, government and communities. A plan for Just Transition provides and guarantees better and decent jobs, social protection, more training opportunities and greater jobs security for all workers affected by the global warming and climate change policies.
>
> International Trade Union Confederation
> (n.d.; in Heyen *et al.* 2020)

In summary, just transition 'describes the transition towards a low-carbon and climate-resilient economy that maximizes the benefits of climate action while minimizing hardships for workers and their communities' (International Trade Union Confederation 2015).

Other stakeholders in this field interpret the concept in a much broader sense, integrating global inequality and environmental concerns into their definition:

> By 'just' we mean: some chance of a safe climate for future generations; an equal distribution of the remaining global carbon budget between countries; and a transition ... in which the costs

are distributed progressively, and where everyone's essential needs for housing, transport and energy use are met.

Friends of the Earth (2011; in Heyen *et al.* 2020)

Here we find complementary keywords like 'future generations' and 'solidarity' between countries.

Nonetheless, some things still remain unclear in this definition: just transition can be interpreted in a very narrow but also in a very broad way. Because of this, some questions can be asked. For example, what is the scope and breadth of the concept? For whom will it be just? What will the transition look like? How will it take place? How will the distributional impacts be allocated proportionately? Different perspectives on just transition assume different meanings when answering these questions, so it will come as no surprise that they can lead to divergent answers. The Just Transition Initiative (2021) developed a preliminary framework to get a better grip on these diverse perspectives, which could help stakeholders to understand the key dimensions of just transitions. Two different axes

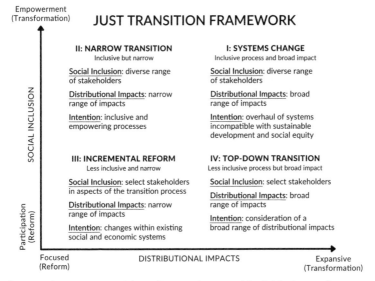

Figure 1. About Just Transitions. (*Source*: justtransitioninitiative.org.)

can be distinguished in this typology: social inclusion and scope (Cahill *et al.* 2020). Action is needed across both dimensions to limit climate-change-related temperature rises to under 2 °C.

Social inclusion indicates the extent to which marginalized groups are included in discussions and decision-making processes (Cahill *et al.* 2020). The interests of people in society are rarely homogeneous, and all their different concerns should be voiced. Social inclusion can range from influencing decision-making processes to some degree to actually challenging unequal power relations and empowering groups that are not usually involved. Social inclusion measures thus assess the breadth of recognition that is given to different stakeholders. This dimension includes both recognition and procedural justice. Recognition points to whose interests are taken into account when formulating new ideas (Svarstad and Benjaminsen 2020). Misrecognition often happens in relation to social categories, such as gender, race, religion and ethnicity. Procedural justice is more about the decision-making process itself. It captures which stakeholders, figures or institutions get the right to decide, participate and ultimately have an influence on the decisions that are made. There is therefore a distinction between legal and intersubjective recognition. Also, it is crucial that the state and the legal system demarginalize these groups in society.

The second dimension – scope – encompasses all the distributional impacts associated with a just transition (Cahill *et al.* 2020). This refers to the fair allocation of both the advantages and the disadvantages of a just transition process. As Walker (2012) emphasized in his 'double injustice' theory, some people are more affected by environmental impacts than others. This should be considered when distributing the burdens and benefits of a just transition process. Both the benefits and the harms of transition outcomes should be considered in terms of future impact. This can range from a focus on the direct impacts (such as job losses in specific sectors) to a broader approach that incorporates a bigger impact across sectors and stakeholders. Together, these two dimensions measure both the impact and the process of just transition.

The scope dimension also involves 'intention' (Cahill *et al.* 2020). This indicates the ideological preference between reforming and transforming the current political and economic system. The important thing here is the pace and scale of the just transition initiatives to realize the necessary changes. On the one hand, 'intention' can indicate the will to achieve a change within the system, which can be market-driven change, creating regulations to provide a 'green economic capitalist system'. At the other end of the spectrum, it can indicate the desire to overhaul the existing system. This requires alternative visions and pathways. The latter option can override other systemic forms of oppression as well, such as racism, sexism, colonialism and classism. One example of such a transformative approach is the focus on degrowth or post-growth (Hickel 2020; Jackson 2009):

> Post-growth scholarship calls for high-income nations to shift away from pursuing GDP growth and to focus instead on provisioning for human needs and well-being, ensuring living wages, shortening the working week to maintain full employment, and guaranteeing universal acces to public healthcare, education, transportation, energy, water and afforbable housing. This approach enables strong social outcomes to be achieved without growth, and creates space for countries to scale down ecological destructive and socially less necessary forms of production and consumption.
>
> Hickel *et al.* (2021)

The framework consists of four different quadrants that all reflect different degrees of social inclusion and scope (Just Transition Framework 2020). The first quadrant pleads for a system change: out of all the quadrants, this is the most far-reaching just transition view in terms of scope and social inclusion. In this scenario, environmental sustainability and social equity are given a lot of importance. Moreover, the way to get there is by local community-led efforts, such as cooperatives. The second quadrant is less far-reaching and more focused in scope, though it is still characterized by a large demand for social inclusion and empowerment. The third quadrant is focused on a relatively targeted set of distributional impacts, such as workers losing their jobs in certain sectors (e.g. in the coal sector). It thus has a narrower focus in comparison with the other quadrants. In addition, only some stakeholders get to participate and engage in the different phases of the decision-making process. The fourth and last quadrant seeks transformative reform on a wider scale but through a top-down transition. This means that only certain stakeholders get to voice the way in which the just transition process should go, so there is minimal social inclusion and participation.

This chapter began with an analysis of the various systemic faults, so it is clear that transformative systemic change is necessary. This book is therefore positioned within the first quadrant, advocating a definition of just transition that emphasizes the need for both structural and transformative reform. A structural reform approach to just transition highlights the following:

> [An] inclusive and equitable decision-making process guiding the transition, and collective ownership and management of the new, decarbonized energy system by the different stakeholders – rather than a single interest (see, for example, McCauley *et al.* 2013). Such an approach to Just Transition implies institutional change and structural evolution of the system. Solutions are not solely produced via market forces or traditional forms of science or technology, but emerge from modified governance structures, democratic participation and decision making, and ownership.
>
> Healy and Barry (2017; in Just Transition
> Research Collaborative 2018)

Structural changes to our dominant societal and economic models have been lacking when it comes to sustainability narratives. With increasing technological advances and innovations, the illusion has been maintained that by swapping energy sources and materials for 'sustainable' alternatives, current consumption patterns can be upheld. In the field of energy transition, for example, quite a few initiatives focus on energy efficiency. Meanwhile, social demands are given less importance. As a result, while increased efficiency and optimization allow industrial tools and vehicles to consume less fossil fuel, a rise in consumption has actually created a steep rise in the use of fossil fuels worldwide. To put it simply: we have more cars and drive further in them.

A transformative approach to just transition, on the other hand, implies

… an overhaul of the existing economic and political system that is seen as responsible for environmental and social crises (Hopwood *et al.* 2005; Healy and Barry 2017). In addition to changing the rules and modes of governance, proponents of this approach also promote alternative development pathways that undermine the dominant economic system built on continuous growth, and imply profoundly different human-environment relations.

Just Transition Research Collaborative (2018)

An example of this approach is provided by Cooperation Jackson, which

sees just transition as fitting within a broader struggle to 'end our systemic dependence on the hydro-carbon industry and the capitalist driven need for endless growth on a planet with limited resources, while creating a new, democratic economy that revolves around sustainable methods of production and distribution that are more localized and cooperatively owned and controlled'.

Just Transition Research Collaborative (2018)

It is, as post-growth thinkers make clear, a shift from satisfying consumer preferences to fulfilling essential human needs by building adequate provisioning systems.

NECESSARY ELEMENTS IN A JUST TRANSITION PROCESS IN PRACTICE

Having argued for the need to integrate just transition as a guiding principle in future trajectories and policies, the obvious next question is how to operationalize this. Insights are also growing in this area. For instance, according to research carried out by Heyen *et al.* (2020), a just transition process must incorporate the following elements.

- *Inclusive social and regional dialogue.* This involves consultations and negotiations between governments and both employers' and workers' associations, as well as environmental, climate and consumer NGOs.

- *Measures to mitigate negative effects on workers and regions.* These include short-term and defensive measures (e.g. compensation for losses) as well as more forward-looking and proactive measures aimed at structural reorientation (e.g. skills development policies and the greening of the economy). Regional eco-development plans seem to be key in the revitalization of regions with high concentrations of 'brown' industries.
- *Support for workers in declining industries to find work.* These are basic measures to facilitate re-employment and they include job placement services, job search training, relocation assistance, recruitment incentives for companies and business formation support.
- *Support for new businesses and the creation of decent work.* Governments should support the creation of new 'green economic activities' with start-up grants, research and development subsidies, etc. Special attention must be paid to creating equal working opportunities for women.
- *Support for specific regions and communities.* National governments can support affected regions and communities by investing in sustainable infrastructure (mobility, energy, etc.) and by relocating public institutions into the regions affected. Top-down support should be coupled with bottom-up processes for the development of visions and ideas for the regions' futures.
- *Enabling all citizens to live sustainable lives.* Financial support for energy-efficient housing renovation, convenient and affordable public transport, and networks for sharing tools and repair services makes it easier for low-income groups to opt for sustainable lifestyles.
- *Protection of vulnerable households from energy poverty.* Measures could include the dedication of carbon tax revenues to helping those most affected by higher energy prices.
- *Support for affected companies.* Business adaptation can be facilitated via, for example, national tax reforms, increasing taxes and duties on resource consumption and emissions while at the same time (and to the same extent) decreasing the tax and duty burdens on the production factor of labour (incidental wage costs).

Moreover, emphasis must be placed on the need both to start as early as possible and to leave room and time for the gradual adjustment of policy plans: 'The earlier that actors have anticipated, accepted and implemented steps to prepare and cushion transition shocks, the better the results.' This needs to be accompanied by 'concrete timelines with clear, consistent middle- and long-term goals' (Heyen *et al.* 2020).

In a world characterized by multi-level governance, just transition initiatives are needed at all levels: at the level of companies and economic sectors, cities, regions and nations, at the EU level and on a global scale. At all these levels, different specific challenges, groups of stakeholders and power relations show that there is no 'one size fits all' model of just transition. It reveals the need to document the different layers, geographic distinctions and specificities of each case.

At national or state level, comprehensive just transition strategies have been developed in the Canadian state of Alberta, among other places. At the regional level, the German Ruhr region provides a relatively successful example of a fundamental transformation from coal and steel to a knowledge-based economy over a time span of 60 years (Heyen *et al.* 2020). And at the company level, Enel, the second-largest electricity company in Europe, with the Italian state as its main shareholder, is planning to close all of its coal- and lignite-fired power plants by around 2030 and has committed to fully decarbonizing its energy mix by 2050 (Smith 2017).

JUST TRANSITION IN THE EU

In the EU, recent reports have focused on the need for transformative politics, echoing the sense of urgency expressed in the IPCC climate and IPBES biodiversity reports. For instance, the first line of the 2020 'State of the environment' report asserts that 'Europe faces environmental challenges of unprecedented scale and urgency' (European Environment Agency 2019). Therefore, the Union 'needs to find ways to transform the key societal systems that drive environment and climate pressures and health impacts – rethinking not just technologies and production processes but also consumption patterns and ways of living. This will require immediate and concerted action, engaging diverse policy areas and actors across society in enabling systemic change.'

In order to achieve these goals, the European Commission (2019) has adopted, in addition to its existing environmental policies, a roadmap for a European Green Deal. The Green Deal aims to achieve climate neutrality in the EU by 2050 with a modern, resource-efficient and competitive economy. Composed of around ten key elements, it articulates strategies aimed at sustainable finance and mobility, as well as a new action plan on the circular economy, a 'Farm to Fork' Strategy, and a Sustainable Europe Investment Plan to provide the necessary funding. Moreover, the EU gives special attention to its newly developed JTM: a funding mechanism composed of three main pillars that was put in place to allocate funding to the regions and sectors most affected by the transition and to ensure that this is carried out in a fair and just manner.

EU climate neutrality and energy policy

As the energy sector is currently responsible for more than 75% of the EU's greenhouse gas emissions (European Commission 2018), it is a crucial sector for climate policy. Transitioning the energy sector offers great opportunities for employment gains. Studies foresee a net employment gain within the EU of 1–2 million jobs by 2030 (Heyen *et al.* 2020). At the same time, going climate neutral will of course have far-reaching effects on the fossil fuel and carbon-intensive sectors. This will have a significant impact on localities and regions where fossil-fuel-dependent sectors are concentrated. An important example is coal mining, which is regionally concentrated, located mostly in Eastern Europe. The steel, cement and chemical industries are also likely to face difficult transitions.

If we look at the lives of citizens, going climate neutral can have huge public health benefits, particularly for vulnerable groups. At the same time, carbon taxes can make life more expensive (e.g. rising prices for electricity and heating). It is therefore important for special attention to be given to low-income households and the unemployed. In this context, the European Commission's Renovation Wave initiative, designed to improve the energy performance of buildings as part of the European Green Deal, could be very influential. The European Commission (2020c) also states that 'given the labour-intensive nature of the building sector, which is largely dominated by local businesses, renovations of buildings also plays a crucial role in the European recovery of the COVID-19 pandemic'. According to their strategy document 'A Renovation Wave for Europe: Greening Our Buildings, Creating Jobs, Improving Lives': 'Today, only 1% of buildings undergo energy efficient renovation every year. A faster rate of renovation is necessary to improve energy efficiency and reduce greenhouse gas emissions.'

The circular economy and resource efficiency

In 2020 the EU introduced a new Circular Economy Action Plan, building on circular economy actions implemented since 2015 (European Commission 2020b). This resulted in different strategies and directives focusing on waste and plastics, among other topics. Importantly, eco-design regulations now focus on the lifespan, maintenance, repair and reuse of products. Although the circular economy is still more a goal than a reality, once implemented it could yield many new jobs. It would also be accompanied by the need for job reallocation, as sectors working with raw materials are likely to decline in size, while the recycling and repair sector could experience substantial growth. Geographical disparities may also increase, with trends such as urban mining creating jobs in more densely populated areas.

Agriculture

The food system accounts for 32% of greenhouse gas emissions globally today. Europe lies around this average, with agriculture accounting for 12% of EU emissions, to which we must add processing, packaging, cold chain energy consumption, inputs for industrial agriculture, and more… Transitioning to agroecology, reducing consumption of meat and dairy products, as well as using less input-hungry production methods, can therefore play a decisive role.

De Schutter (2018)

A number of studies indicate that the transition to sustainable agriculture could create many new full-time jobs across the entire food production chain, as low-impact (organic) farming methods are more labour intensive than conventional, industrial farming. As such, a 2006 study conducted by the Soil Association concluded that organic farms in the UK provided 32% more jobs per farm than conventional farms.

The Just Transition Mechanism

The interpretation of just transition adopted by EU policy tends to fit into a more traditional green growth narrative. Promoting economic growth remains crucial, and it should not be considered to be incompatible with environmental protection and social progress. Sabato and Fronteddu (2020) evaluate the goals of the European Commission President, Ursula von der Leyen:

> The fight against climate change indeed features strongly in the political programme of the new Commission President Ursula von der Leyen. In her Political Guidelines for the Next Commission, she stated that the key priority of her Commission would be to transform Europe into '… the first climate-neutral continent' (von der Leyen 2019, p. 5), by developing a European Green Deal (EGD). In so doing, one of the priorities should be to ensure a 'just transition for all' (ibid. 6).
>
> Sabato and Fronteddu (2020)

The European Green Deal should serve as 'a new growth strategy that aims to transform the EU into a fair and prosperous society, with a modern, resource-efficient and competitive economy where there are no net emissions of greenhouse gases (GHG) in 2050 and where economic growth is decoupled from resource use' (European Commission 2019, p. 2; cited in Sabato and Fronteddu 2020).

The policies and measures necessary to achieve the ambitious targets of the European Green Deal may imply a negative impact on jobs and regional economies, especially in the regions that depend on fossil fuel and carbon-intensive industries. Furthermore, to achieve carbon neutrality, financial investment is needed. In order to support those affected, the European Commission proposed the creation of a Just Transition Fund in January 2020 as part of its Green Deal. This fund is available to all member states, but its biggest focus is on the regions potentially most affected by transition policies, where the biggest changes will have to take place. The budget proposed by the European Commission amounts to €40 billion, which should be complemented with resources from both cohesion policy funds and national co-financing. The fund will be part of the JTM (European Commission 2020a). The two other pillars of the JTM are the InvestEU Just Transition scheme and the European Investment Bank public sector loan facility, which will bring in private investment and public funding, respectively. The JTM's total budget is expected to reach €100 billion (Widuto and Jourde 2020).

But how much of the budget will go to EU member states? This is assessed based on a range of social and economic criteria, including employment and GHG emissions in the regions. Member states will receive access to the fund by making territorial just transition plans, wherein the most impacted regions are identified. In this scheme, Germany and Poland will benefit most: two-thirds of coal mining in Europe takes place in Poland and half of the coal-dependent jobs are situated there (Galgóczi 2018). At the moment, however, Poland is nowhere near to planning to phase out coal (Popp and de Pous 2020).

THE EU'S JUST TRANSITION: CHALLENGES AND OPPORTUNITIES

According to Francesca Colli from the Egmont Institute (2019), the JTM, which 'foresees mobilising at least €100 billion through a combination of public and private investments', has three main targets: 'people and citizens most vulnerable to the transition; companies and sectors active in carbon-intensive industries; and Member States and regions that are dependent on fossil fuels and carbon-intensive industries'. She has identified three main challenges that the JTM needs to address in order to achieve its objectives.

- *Moving the focus away from national allocations.* One pillar of the JTM, the Just Transition Fund, has been subject to criticism as it draws from the EU's cohesion funds, and this has fuelled divisions between (groups of) countries.
- *Moving beyond energy production.* 'A just transition is not only about energy production and more obvious carbon-intensive industries, but also about systemic change ... necessary in many sectors ... that will affect workers, consumers and citizens.' Other sectors in need of transition include transport, construction and renovation.
- *Working with the private sector and stakeholders.* Companies are a fundamental actor to have on board during the transition. It is therefore necessary for the EU to ensure that concrete regulations exist, at both the national and EU levels, such as targets and timelines for the phasing out of fossil fuels. 'It is [also] important to keep in mind the "polluter pays" principle ... [which] is not used efficiently when used to subsidise or resolve the results of negative externalities of companies' activities.' Moreover, there are not enough measures that require the consultation and inclusion of stakeholders at different levels. Further investigation and reassessments are therefore needed.

Although these three points are very valid, they are not the main issue at stake regarding the EU policies on climate change and just transition. In light of the new IPCC report, whose content was leaked in the summer of 2021, it is clear that the EU policies coined as 'Fit for 55' show more ambition than what was envisioned before, but they are still really not enough. This means that there is an increasing urge for fast and effective measures, which in turn increases the risk that communities will be left behind if we fail to ground all policies in the principle of a just transition.

Bibliography

Alexander, D. 2010. The impact of the economic crisis on the world's poorest countries. *Global Policy* 1(1), 118–120 (doi: 10.1111/j.1758-58 99.2009.00018.x).

Armelagos, G. J., and Harper, K. N. 2009. Emerging infectious diseases, urbanization, and globalization in the time of global warming. In *The New Blackwell Companion to Medical Sociology*, edited by Cockerham, W. C. Wiley.

Bollen, Y., De Wel, B., and Verschoore, V. 2019. De klimaatontwrichting als vakbondsthema: een rechtvaardige transitie voor werknemers. In *Klimaat en sociale rechtvaardigheid*, edited by Dierckx, S, pp. 111–134. Antwerp, Belgium: Gompel&Svacina (In Dutch).

Brand, U. 2016. How to get out of the multiple crises? Contours of a critical theory of social-ecological transformation. *Environmental Values* 25(5), 503–525 (doi: 10.3197/096327116X14703858759017).

Cahill, B., Margaret, M., and Landislaw, S. 2020. Just transition concepts and relevance for climate action. Report, Center for Strategic & International Studies.

Colli, F. 2019. The EU's just transition: three challenges and how to overcome them. European Policy Brief, No. 9, Egmont Royal Institute for International Relations (https://bit.ly/3Bwz2HG).

De Schutter, O. 2018. System overhaul: making food sustainable. *Green European Journal*, 21 November (https://bit.ly/3gZR8bH).

European Commission. 2018. A clean planet for all: a European strategic long-term vision for a prosperous, modern, competitive and climate neutral economy. Report, November, Brussels (https://bit.ly/36dTsJS).

European Commission. 2019. A European Green Deal – striving to be the first climate-neutral continent. Report, Brussels (https://bit.ly/3HinSIK).

European Commission. 2020a. Allocation method for the just transition fund. Report, 15 January, Brussels (https://bit.ly/3H5zjlT).

European Commission. 2020b. A new circular economy action plan: for a cleaner and more competitive Europe. Report, 11 March, Brussels. (https://bit.ly/35CQBKm).

European Commission. 2020c. A renovation wave for Europe – greening our buildings, creating jobs, improving lives. Report, Brussels (https://bit.ly/3uXBG8b).

European Environment Agency. 2019. The European environment – state and outlook 2020: knowledge for transition to a sustainable Europe. Report, Publications Office of the European Union, Luxembourg (www.eea.europa.eu/publications/soer-2020).

Galgóczi, B. 2018. Just transition towards environmentally sustainable economies and societies for all. ILO ACTRAV Policy Brief, International Labour Organization, Geneva (https://bit.ly/3I8MKmM).

Healy, N., and Barry, J. 2017. Politicizing energy justice and energy system transitions: fossil fuel divestment and a 'just transition'. *Energy Policy* 108, 451–459.

Heyen, D. A., Menzemer, L., Wolff, F., Beznea, A., and Williams, R. 2020. Just transition in the context of EU environmental policy and the European Green Deal. Report, March 2020, Öko-Institut, Freiburg (https://bit.ly/3uW0mxT).

Hickel, J. 2020. *Less Is More: How Degrowth Will Save the World*. London: Cornerstone/Penguin Books.

Hickel, J., Brockway, P., Kallis, G., Keyßer, L., Lenzen, M., Slameršak, A., Steinberger, J., and Ürge-Vorsatz, D. 2021. Urgent need for postgrowth climate mitigation scenarios. *Nature Energy* 6, 766–768 (doi: 10.1038/s41560-021-00884-9).

Holemans, D. 2021. Freedom & security in a complex world. Essay, Green European Foundation (https://bit.ly/3gUn2WX).

Hopwood, B., Mellor, M., and O'Brien, G. 2005. Sustainable development: mapping different approaches. *Sustainable Development* 13, 38–52.

Intergovernmental Panel on Climate Change. 2018. Global warming of 1.5 °C. An IPCC special report on the impacts of global warming of 1.5 °C above pre-industrial levels and related global greenhouse gas emission pathways, in the context of strengthening the global response to the threat of climate change, sustainable development, and efforts to eradicate poverty. (www.ipcc.ch/sr15/).

Intergovernmental Science-Policy Platform on Biodiversity and Ecosystem Services. 2019. Global assessment report on biodiversity and ecosystem services (www.ipbes.net/global-assessment).

International Trade Union Confederation. 2015. Climate justice: there are no jobs on a dead planet. Frontlines Briefing, March (https://bit.ly/3LMaDCw).

Jackson, T. 2009. *Prosperity Without Growth: Economics for a Finite Planet*. Abingdon: Routledge.

Just Transition Initiative. 2021. About just transitions. Website (https:// justtransitioninitiative.org/).

Just Transition Research Collaborative. 2018. Mapping just transition(s) to a low-carbon world. Report, UNRISD (https://bit.ly/3s1Fi64).

Koch, M. 2018. Sustainable welfare, degrowth and eco-social policies in Europe. In *Social Policy in the European Union: State of Play 2018*, edited by Vanhercke, B., Ghailani, D., and Sabato, S., p. 35. Nineteenth Annual Report of the European Trade Union Institute, Brussels (https://bit.ly/3rdPQj2).

Labor Network for Sustainability. n.d. 'Just Transition' – just what is it? An analysis of language, strategies, and projects. Website (https://bit.ly/3rVFy7D).

Mazzocchi, T. 1993. An answer to the jobs–environment conflict? *Green Left Weekly* 114, 8 September (https://bit.ly/3GZ1rr6).

McCauley, D. A., Heffron, R. J., Stephan, H., and Jenkins, K. 2013. Advancing energy justice: the triumvirate of tenets. *International Energy Law Review* 32(3), 107–110.

Milanovic, B. 2017. Wereldwijde ongelijkheid: Welvaart in de 21e eeuw. Houten: Spectrum, pp. 21–60 (In Dutch).

Organisation for Economic Co-operation and Development. 2019. Beyond growth: towards a new economic approach. Report of the Secretary General's Advisory Group on a New Growth Narrative, OECD (https://bit.ly/3GYka5Z).

Piketty, T. 2013. *Le Capital au XXIe Siècle*. Paris: Seuil (In French).

Polanyi, K., and MacIver, R. M. 1944. *The Great Transformation*, Volume 2, p. 145. Boston, MA: Beacon Press.

Popp, R., and de Pous, P. 2020. Just transition fund can boost European coal phase-out. *Social Europe*, 17 February (https://bit.ly/36dTJfS).

Powell, D., Balata, F., Van Lerven, F., and Welsh, M. 2019. Trust in transition. Report, 28 November, New Economics Foundation (https://bit.ly/35aMrct).

Sabato, S., and Fronteddu, B. 2020. A socially just transition through the European Green Deal? Working Paper, ETUI, Brussels (https://bit.ly/3I2CZ9y).

Smith, S. 2017. Just transition: a report for the OECD. Report, May, Just Transition Centre, Brussels (https://bit.ly/3v4f85B).

Steffen, W., Richardson, K., Rockström, J., Cornell, S. E., Fetzer, I., Bennett, E. M., and Sorlin, S. 2015. Planetary boundaries: guiding human development on a changing planet. *Science* 347(6223), 736–747 (doi: 10.1126/science.1259855).

Svarstad, H., and Benjaminsen, T. A. 2020. Reading radical environmental justice through a political ecology lens. *Geoforum* 108, 1–11.

Tsing, A. L. 2008. Contingent commodities. In *Taking Southeast Asia to Market*, edited by Nevins, J., and Lee Peluso, N. Cornell University Press.

Tsing, A. L. 2011. *Friction: An Ethnography of Global Connection*. Princeton University Press.

Tsing, A. L. 2015. *The Mushroom at the End of the World*. Princeton University Press.

Union to Union. 2020. Just transition in the international development cooperation context. Report, October (https://bit.ly/3LFDEzU).

United Nations Research Institute for Social Development. 2018. Mapping just transition(s) to a low-carbon world. Research Report, 28 November, UNRISD, p. 6 (https://bit.ly/3wstIV6).

United Nations Research Institute for Social Development. 2021. A new eco-social contract: vital to deliver the 2030 agenda for sustainable development. Research Report, March, UNRISD 11, pp. 1–4.

van de Pas, R. 2020. Globalization paradox and the coronavirus pandemic. Report, May, Netherlands Institute of International Relations, Clingendael (https://bit.ly/3gZRGhL).

Vidal, J. 2020. Destroyed habitat creates the perfect conditions for coronavirus to emerge. Blog Post, *Ensia* (https://bit.ly/3v4sAq6).

Walker, G. 2012. *Environmental Justice: Concepts, Evidence and Politics*. London: Routledge.

Widuto, A., and Jourde, P. 2020. Just transition fund. Briefing: EU Legislation in Progress 2021–2027 MFF. European Parliamentary Research Service, Brussels (https://bit.ly/3rZc88H).

A global perspective: voices from the South

Daniel Chavez, Rand El Zein, Lyda Fernanda Forero and Anya Namaqua Links

INTRODUCTION
By Daniel Chavez

In recent years, the concept of a *just transition* has become a recurrent buzzword in the discourse of political, economic and social actors from diverse geographical, institutional and ideological backgrounds. The concept began to grow in popularity six years ago, when the preamble to the Paris Agreement acknowledged the need for 'a just transition of the workforce and the creation of decent work and quality jobs in accordance with nationally defined development priorities' (21st Conference of the Parties 2015). Similar aims were also included in the wording of the Sustainable Development Goals (United Nations 2015), and they were ratified by the Solidarity and Just Transition Silesia Declaration, signed by fifty countries at the 24th Conference of the Parties (COP24) in 2018.

Despite the rapid and widespread dissemination of the concept, it risks becoming another empty signifier. The limited debate around its content has increasingly turned it into an ahistorical and apolitical idea. In fact, the apparent consensus around its meaning showcases that insufficient debate has taken place on the many and distinct *transitions* (plural) that are needed in different national contexts, as well as the clear links to deep-rooted relationships of exploitation, domination and environmental plunder that persist around the world. There is an obvious need for a deeper discussion around the concrete significance and scope of this term, especially in those countries most affected by climate change and hit hardest by the Covid-19 pandemic.

Long before references to a just transition appeared in official documents, a wide range of social movements had already denounced growing inequality and the extreme spoliation of both nature and human beings. They had given very concrete proposals to advance towards social and climate justice. More recently, academia has also included the notion of just transition in the research agenda of scholars from multiple disciplines.

A quick review of the institutional, technical and academic literature, as well as recent reports produced by social and environmental organizations, indicates that the notion of a just transition is linked to virtually every segment of the economy: from agriculture to food production, and from the financial system to energy generation, distribution and consumption. Energy, in particular, is the area that has attracted most of the attention among just transition advocates, given its links to other areas of the global economy and its impact on climate change.

Current exchanges around just transition have developed in parallel with thought-provoking discussions around the concepts of *energy justice* (and more recently around the ideas of *energy democracy* and *energy sovereignty*) and *climate justice*. The former refers to issues of equity, looking at winners and losers within the processes and structures of energy production and consumption, with emphasis on the following: the goals and impacts of energy policies, the participation of workers and users, and the recognition of the rights and demands of local communities affected by energy projects (Jenkins *et al.* 2016). The latter addresses questions of historical responsibilities, the meaning of 'development', and the significance and scope of human and environmental rights. In short, it implies 'a transition path that reconciles the material needs of the poorest people on the planet with the need to safeguard the stability of the Earth's climate', as well as the necessity to simultaneously address 'the overuse of natural resources and the under-provision of public goods' (Jacob and Steckel 2016, p. 2).

The analyses and proposals around just transition in general, and the energy transition in particular, reflect disparate and even antagonistic interests. On the one hand, there are neoliberal visions that propose a fast shift to a sort of 'green capitalism' and that emphasize the centrality of the market as the best or only way to initiate and catalyse transition. On the other hand, there are neo-Keynesian or Marxist views, arguing that strong state intervention is an essential condition for any transition. In relation to the explicit goals, it is easy to distinguish a plurality of perspectives, ranging from a blind faith in solving our current and future problems with technology to proposals around 'degrowth' (Hickel 2020) and 'environmentalism of the poor' (Anguelovski and Martinez Alier 2014) that revalue local community-based knowledge and practices. The idea of just transition has also been applauded and supported by, on the one hand, large energy corporations that perceive new opportunities for financial resilience and profit growth by moving away from fossil fuels and by, on the other hand, local energy communities and progressive local governments.

A primer on the energy transition, jointly produced in 2020 by the activist research centres Taller Ecologista (based in Argentina) and the Transnational Institute (TNI, based in Amsterdam), proposed a distinction between two understandings of *transition*. The first approach, which has been characterized as the *corporate transition*, derives from a techno-economic perspective that focuses on lowering greenhouse gas emissions, 'within a growing process of accumulation of wealth and power through new extraction areas, maintaining existing power relations and therefore also inequality' (p. 2). The second approach, categorized as a *people's energy transition*, 'is based on the premise of constructing the right to energy and questions the idea of energy as a commodity' (p. 3). This perspective implies the de-privatization of the energy system and the strengthening of diverse forms of public ownership, with increasing levels of citizen and worker participation and democracy within the energy system.

Yet, the corporate transition is not advocated only by private or public profit-focused corporations. It is also defended by many other institutions, with varying degrees of enthusiasm or support, including national and local governments and even some influential environmental non-governmental organizations (NGOs) that perceive market forces as the only possible pathway – or, for some, the 'fastest' pathway – for tackling the spread and the urgency of the climate crisis.

Clearly, the issues of ownership and the control of access to energy sources, materials and technologies are at the core of current discussions around the meaning of – and the possibilities for – an energy transition. Many advocates of the market-driven transition promote the expansion of renewable energy sources from a purely utilitarian and extractivist perspective. Quite often, the concept of 'energy efficiency' is incorporated into this logic, reducing the proposed alternatives for the transition to incremental technological gains and/or individual behavioural changes; these do not substantially challenge the patterns of consumption that are intrinsic to the current structure of the energy system.

A rather less optimistic and more sober evaluation of ongoing transformations in the global energy system has been proposed by researchers and activists linked to the Trade Unions for Energy Democracy (TUED) initiative. Preparing for COP26 in Glasgow (November 2021), the TUED and the TNI jointly published a report, as a contribution to exchanges among environmental and labour organizations, in which the authors confronted the failures of neoliberal climate policy. They argued that *the energy transition remains a myth* and that the world needs to overcome its current denialism and recognize the growing gap between ambition and action (Sweeney *et al.* 2021).

Advocates of the market, pointing to the allegedly 'unstoppable' energy transition, frequently issue assurances that their approach is producing positive results. Business leaders, government officials and representatives of major environmental NGOs often state that 'the problem' is that the pace of the transition is too slow, and they call for more 'ambition', more 'political will' and greater 'urgency'. What they mean by this is the following: private investors must be given more incentives; carbon pricing schemes must proliferate and become more robust in their impact on polluters; subsidies for fossil fuels must be removed as quickly as possible; and energy market liberalization and privatization must be pursued more aggressively than ever. Because the need for climate action is so pressing, they also propose using public funds to further 'leverage', 'unlock' and 'de-risk' private investment, so that new markets can be created and new industries can begin to flourish (Sweeney *et al.* 2021).

According to the most recent report published by the International Energy Agency (2021), in order to achieve the goal of net zero emissions by 2050, the exploitation and development of new oil and gas fields must stop *right now,* and no new coal-fired power stations can be built anywhere in the world. This report and many others demonstrate that despite some changes in the fuel mix in some locations and sectors, overall energy demand has continued to rise even faster than the deployment of new, 'clean' energy sources. As a result, nearly all forms of energy have grown alongside each other. Renewable sources have been a significant contributor to that overall growth – particularly in the power sector – but this has not resulted in any significant displacement of fossil-based energy (Sweeney *et al.* 2021).

In short, the claims that the transition is 'happening' and 'unstoppable' are not supported by the facts. A radical policy shift is necessary – one built on the notion of *global public goods* and the expansion of both direct *public investment* and *public ownership* of key components of the economy, particularly the energy sector. Resistance to privatization has already turned into a powerful force for reclaiming public ownership of energy and other essential services, with thousands of successful (re-)municipalizations and (re)nationalizations having taken place in countries around the world (Kishimoto *et al.* 2020).

At the same time, people are actively engaged in the shift towards renewable sources worldwide, and in challenging the big corporate interests of the old regime. Around the world, we can witness the extension of decentralized forms of ownership structured around citizens, who are pooling resources and capacities to run renewable energy cooperatives and municipal utilities. This has created inspiring results in terms of democracy, community empowerment and local development. However, many issues remain unresolved, because empirical evidence from around the world points to the fact that local and community energy initiatives will not be enough to disrupt the hegemonic for-profit energy model. How will social and public ownership be designed and integrated at the local, regional and national levels? What structures and functions should be owned and run by the state or by local communities? These are deep, vital questions that must be addressed.

Bibliography

Anguelovski, I., and Martínez Alier, J. 2014. The 'Environmentalism of the Poor' revisited: territory and place in disconnected glocal struggles. *Ecological Economics* 102, 167–176.

Hickel, J. 2020. What does degrowth mean? A few points of clarification. *Globalizations* 18(7784), 1–7.

International Energy Agency. 2021. Net zero by 2050. A roadmap for the global energy sector. Report, IEA, Paris (https://bit.ly/36qmRRF).

Jakob, M., and Steckel, J. C. 2016. The just energy transition. Working Paper, supported by WWF International (https://bit.ly/3iyrXO4).

Jenkins, K., McCauley, D., Heffron, R., Stephan, H., and Rehner, R. 2016. Energy justice: a conceptual review. *Energy Research & Social Science* 11, 174–182.

Kishimoto, S., Steinfort, L., and Petitjean, O. 2020. *The Future is Public: Towards Democratic Ownership of Public Services.* Amsterdam: Transnational Institute.

Sweeney, S., Treat, J., and Chavez, D. 2021. Energy transition or energy expansion? Report, 22 October, Transnational Institute (TNI) and Trade Unions for Energy Democracy (TUED) (https://bit.ly/3Nm5zG6).

Transnational Institute and Taller Ecologista. 2020. Transiciones energéticas. Aportes para la reflexión colectiva. Report, September, Transnational Institute and Taller Ecologista (https://bit.ly/3wzasFQ) (In Spanish).

JUST TRANSITION AND DEVELOPMENT: A BATTLE AND CHALLENGE FOR NAMIBIA
By Anya Namaqua Links

The situation

Namibia is a sparsely populated, drought-prone country in south-western Africa. Once a colony under German and then South African administration, it gained independence in 1990 during the so-called 'third-wave of democratization' that swept across sub-Saharan Africa between 1989 and 1995.

After independence, Namibia became a constitutional democracy and adopted a mixed economic system where centralized economic planning and government regulation coexisted with free-market elements such as private ownership and profit-seeking, while strategic industries providing public goods were converted into state-owned enterprises (SOEs).

Namibia and its larger neighbour South Africa currently rank as two of the most unequal countries in the world, characterized by severe disparities in standards of living and income. Although the World Bank has classified Namibia as an upper-middle-income country, it has a high Gini index: 57.6 in 2020.

As a combined consequence of underdevelopment and low industrialization during apartheid and colonialism, Namibia's secondary sector is poorly developed; the country net imports more than 60% of the goods consumed by the domestic market – such as energy (fuels and electricity), machinery, vehicles and cereals – while it exports commodities such as copper, diamonds, uranium and marine resources to generate foreign income.

In 2004, the Namibian government launched Vision 2030: a high-level policy framework for long-term national development to eradicate poverty, unemployment and inequality by the year 2030. Although the country's GDP grew considerably between 1995 and 2020, that growth was not accompanied by development, so it did not translate into improved living conditions for most Namibians. According to the World Bank, approximately 1.6 million Namibians (64%), out of a population of 2.5 million, live below the upper-middle income poverty line of $5.50 per day: higher than the average poverty headcount ratio of 40.4% for sub-Saharan Africa.

Structural unemployment, a lack of job opportunities, a shortage of skilled workers and a 'brain drain' characterize the Namibian labour market. The national unemployment rate stubbornly remains higher than 30%, with youth unemployment (for those aged 15–34) at 46%: significantly higher than the national average. The largest employer in Namibia, the agricultural sector, is also the lowest-paying sector overall. In urban areas, 57% of the labour force are employed in the informal economy and slightly more than 40% are employed in the formal economy. Labour unions are uncoordinated and fractured into industries, and unions primarily focus on members keeping their jobs.

Since gaining independence in 1990, Namibia has consistently invested in a national development budget that targeted its human capital via education and health. For example, school feeding programmes were implemented. Primary education in Namibia is compulsory and free, resulting in an adult literacy rate of 92%. The country's health expenditure is (with South Africa's) the highest on the African continent, while it is one of only seven countries in sub-Saharan Africa that has in place social safety nets such as food banks and grants for the disabled, children and the elderly. Most Namibians currently live within a 10 kilometre radius of a health care facility. Life expectancy at birth (over a thirty-year period since independence) has increased to 65, higher than the sub-Saharan average of 62. In addition, 77% of those Nambians who are living with HIV (12% of the population) are virally suppressed due to national antiretroviral therapy programmes. Overall, Namibia reduced its Gini income poverty and inequality index from 63 in 2003 to 57.6 in 2020 (an index of 100 being indicative of perfect inequality).

Namibia's 'triple challenge' – poverty, unemployment and inequality (PUI) – has three distinct dimensions: namely, an urban–rural divide; gender; and finally race, ethnicity and class. The urban–rural divide is evident throughout the country: essential services, resources and employment opportunities are concentrated in urban areas, whereas 49% of the population lives in poorly serviced, underdeveloped rural areas. The Ministry of Mines and Energy of Namibia estimates that 80% of rural households do not have access to electricity. At 4.5%, Namibia's urban–rural migration rate is higher than its economic and population growth rates, in part because urban household incomes exceed rural household incomes by more than 100%. In terms of gender, 43.9% of households severely affected by PUI are female-headed households, with more men employed across all sectors of the economy except for domestic work. In terms of race, ethnicity and class, the wealthiest households in Namibia are white German and English speaking, and the poorest households are speakers of the indigenous Rukwangali, Khoekhoegowab and Saan (also known as 'Khoisan') languages.

In 2016, Namibia went into economic recession. Despite growth in GDP, various economic initiatives and interventions spearheaded by the government failed to diversify and industrialize the economy and failed to create employment and opportunities or to reduce PUI. A recent study of mean annual precipitation data from 2010 to 2020 showed that Namibia experienced long-term drought for a period of 10 years due to declines in rainfall, with devastating effects on farming and agriculture. In its latest country brief, the World Food Programme reported that 428,000 (or 17%) of Namibians were food insecure, and nationwide 24% of children aged under 5 were stunted because of poorly diversified diets, i.e. malnutrition.

Namibia is a Non-Annex I Party to the United Nations Framework Convention on Climate Change (UNFCCC), defined as a developing country that is 'especially vulnerable to the adverse impacts of climate change' with special needs in terms of investment, insurance and technology transfer. As such, it does not have any commitments under the Convention. In 2015, and again in 2021, Namibia compiled intended nationally determined contributions (INDCs) that outlined its circumstances, a summary of its needs and its climate change adaptation and mitigation contributions. In 2015, financial assistance to the value of $22.6 billion was required to implement INDC components.

The country's president, Dr H. G. Geingob – faced with persistent economic hardship and rising poverty levels that were recently aggravated by the loss of approximately 40,000 jobs during the Covid-19 pandemic – recently launched the Harambee Prosperity Plan (HPP II, 2021–2025): an action plan with economic recovery and inclusive growth as its overarching goals.

More than a just transition

At its core, just transition is concerned with social equity and sustainable development; however, the path to achieving those objectives is bifurcated. Just transition is a directive to either reform or transform existing economic and political systems to avoid a looming climate change crisis and to protect the biosphere. If the intention is to reform existing systems, the implication is that they remain in place but are subjected to reformation through amended guidelines, standards, policy interventions and legislation. If the intention is to transform existing systems, a radical overhaul of such systems is implied. The choice to reform or transform depends on the countries themselves.

The biggest problems with just transition in Namibia are legacy systems from apartheid and colonialism, specifically the long-lasting impact of the 'settler economy', where a marginalized and oppressed indigenous majority served as a source of cheap labour for European settler economic activities in the mining, farming and fishing industries. The apartheid economy exploited the black majority and used legislation and violence to keep indigenous groups 'apart' from the white minority in poorly developed homelands (Namibia had ten such homelands under apartheid). The prolonged lack of access to quality education and opportunities for advancement entrenched poverty, unemployment and inequality for the black majority, and it continues to hamper development in Namibia to this day.

Like other former colonies in sub-Saharan Africa, Namibia's economic structure, organization and existing trade relationships are critical aspects of its colonial legacy and identity. It is important to keep in mind the purpose and function of a colony in relation to the economy of the colonizing country, which makes it difficult for former colonies, and developing countries in general, to unshackle themselves from their colonial economic roles as providers of cheap labour, profit and raw materials.

Because of its status as a former colony of Germany and apartheid South Africa, Namibia's economy is characterized by specialization in the primary sector, severe infrastructure underdevelopment and social inequalities. Unsurprisingly, given its colonial history, the country provides raw materials for innovation and production to take place elsewhere, and is it also a consumer and recipient of finished goods and innovations from elsewhere.

The abovementioned situation also applies to the just transition discourse, and to similar transitions such as just energy transitions, in other former colonies. The decision to transition, including discussions about the necessity of transitioning, typically take place elsewhere and are communicated as well-meaning 'guidelines' to countries like Namibia, without considering the developmental challenges that these countries face: high and entrenched unemployment, for one.

Namibia – like other former colonies with dependent, lagging economies, high unemployment, a poorly educated and unskilled workforce and budget restrictions – has development plans and priorities that may appear counterintuitive to outsiders. Former colonies typically have massive infrastructure deficits when it comes to things like schools, tertiary institutions, health facilities, roads, railways, airports, ports, energy supply, poorly developed public transport systems and housing, to name but a few.

Public expenditure on tangibles – such as food security, drought relief, education, housing and, recently, Covid-19 facilities and vaccines – provides immediate relief and delivers results to affected communities in the short to medium term; however, given the pressing nature of the hardships faced by most Namibians on a daily basis, achieving climate objectives is a long-term and, for the most part, abstract goal.

To accelerate development, and to avoid political instability, the governments of many former colonies have turned to the China Development Bank to procure cheaper, more flexible loans using their oil, mineral, ore, metal and marine resources as security, and they have also sought Chinese expertise on infrastructure development. Shortly after independence, bilateral agreements between China and the Namibian government appointed China as Namibia's official infrastructure development partner. The Chinese government owns uranium mines in Namibia to bolster its own energy agenda, and to date it has received more than 100 infrastructure contracts from the Namibian government, including the expansion of the main port at Walvis Bay.

In conclusion, Namibia has gained substantially by investing in health and education since independence in 1990, and these gains are reflected in its human development index score of 0.646 in 2019 (medium category). However, severe inequality, unemployment and a lack of innovation, skills and job opportunities continue to plague the Namibian labour market and curtail the country's development.

The largest contributor to Namibia's GDP remains its primary sector, i.e. mining, fishing and agriculture, in keeping with its colonial role as a provider of raw materials for resource-poor countries. In February 2022, French supermajor TotalEnergies announced that it had discovered oil in Namibia's offshore Orange Basin: an estimated

11 billion barrels. In response, the Namibian government said that it hoped the discovery of oil would fast-track national development. Given all the above, it is therefore unsurprising that serious discussion about just transition is not on the national agenda, and neither is it part of the country's national development plans.

Bibliography

Centre for Strategic and International Studies and Climate Investment Funds. 2020. Just transition concepts and relevance for climate action: a preliminary framework. Report, 26 June, Just Transition Initiative (https://bit.ly/35iBIMR).

Government of the Republic of Namibia. 2015. Intended nationally determined contributions (INDC) of the Republic of Namibia to the United Nations Framework Convention on Climate Change. Policy Document, September, Windhoek, Namibia (https://bit.ly/3IeGg5y).

Halsey, R. 2018. Reflection paper: considering renewable energy jobs within a just energy transition in South Africa. Website, Project 90 by 2030, Cape Town, South Africa (https://bit.ly/3uefpRu).

Hirsch, T., Matthess, M., and Funfgelt, J. 2017. Guiding principles and lessons learnt for a just energy transition in the Global South. Study, Friedrich Ebert Stiftung (FES), Berlin (https://bit.ly/3LTIM3q).

International Labour Organization. 2015. Guidelines for a just transition towards environmentally sustainable economies and societies for all. Guidelines, ILO, Geneva (https://bit.ly/3sbXmM5).

Jauch, H. 2012. Poverty, unemployment, and inequality in Namibia. *Theme On The Environment, Macroeconomics, Trade And Investment-Series (TEMTI) of Economic Perspectives on Global Sustainability*, February (https://bit.ly/3H9uTe6).

Links, A. 2021. Long-term drought? MAP data for Windhoek show a disturbing trend. Blog Post, 4 January, Euro-African Life in Namibia (https://bit.ly/3t06GBS).

Melber, H. 2007. *Transitions in Namibia: Which Changes for Whom?* Uppsala, Sweden: Nordiska Afrikainstitutet.

Morena, E., Krause, D., and Stevis, D. 2020. *Just Transitions: Social Justice in the Shift towards a Low-Carbon World*. London: Pluto Press.

Namibia Statistics Agency. 2018. The Namibia Labour Force Survey (NLFS). Report, Windhoek, Namibia (https://bit.ly/3ve09ql).

National Planning Commission of Namibia. 2004. Namibia vision 2030: policy framework for long-term national development (summary). Policy Document, Office of the President, Government of the Republic of Namibia (https://bit.ly/3IcpZOG).

National Planning Commission of Namibia. 2017. Namibia's 5th national development plan (NDP5): working together towards prosperity 2017/18–2021/22. Policy Document, NPC, Windhoek, Namibia (https://bit.ly/3s9sDz6).

Ndjavera, M. 2021. Youth unemployment expected to reach 50%. *New Era*, 26 July, Windhoek, Namibia (https://bit.ly/37ULnuB).

Office of the Prime Minister. 2021. The Harambee prosperity plan II: 2021–2025. Policy Document, Government of the Republic of Namibia (https://hppii.gov.na/).

Overy, N. 2018. The role of ownership in a just energy transition. Website, Project 90 by 2030, Cape Town, South Africa (www.90by2030.org.za).

Rama, M., Pursiheimo, E., Lindroos, T., and Koponen, K. 2013. Development of Namibian energy sector. Research Report, VTT Technical Research Centre of Finland (https://bit.ly/3BGtla5).

Sunde, T., and Akanbi, O. A. 2015. Sources of unemployment in Namibia: an application of the structural VAR approach. Report, 12 January, Namibia University of Science and Technology (NUST), University of South Africa (UNISA), Munich Personal RePEc Archive, MPRA Paper 86578: (https://bit.ly/3NjzAq2).

United Nations Development Programme. 2020. The next frontier: human development and the anthropocene. Briefing note for countries: Namibia. Human Development Report, UNDP (https://bit.ly/3Iefkmx).

United Nations Framework Convention on Climate Change. 2014. Parties and observers: Namibia. Website, United Nations Climate Change (https://unfccc.int/parties-observers).

United Nations World Food Programme. 2021. World Food Programme: Namibia (www.wfp.org/countries/namibia).

World Bank. 2020. Literacy rate, adult total (% of people ages 15 and above) – Namibia. Statistics, UNESCO Institute for Statistics (https://bit.ly/3p6ixNq).

World Bank. 2021. The World Bank in Namibia: country overview. Website (https://bit.ly/3v4mqq5).

HEIGHTENED VULNERABILITIES IN LEBANON AND SYRIA AND THE EFFECT OF CLIMATE CHANGE
By Rand El Zein

On 14 October 2019, during a period of high temperatures and strong winds, the dense forests of Mount Lebanon were engulfed by wildfires. The flames reached residential areas, burning houses, schools and small businesses; hundreds of families were displaced and at least three civilians were killed. Not only did the Lebanese government fail to offer the displaced families shelter, nor compensation for their losses, they were also unable to contain the wildfires. Three fire-fighting helicopters were donated to Lebanon in 2009; however, due to the government's incompetence in funding and maintaining them, the helicopters fell into disrepair, with the result that the wildfires spread across the country.

Within a few days of these events, thousands of people filled the streets of Beirut to protest against government corruption and to demand the abolition of the Lebanese sectarian state system.[*] The protests condemned the absence of public services as well as the Lebanese regime's environmental abuses, such as 'mismanagement of waste services that grew into a nation-wide crisis' and 'the illegal quarrying of mountains to nurture the real estate boom' (Fregonese 2019). These abuses resulted in 'higher risks of mudslides and flash floods, patchy planning and zoning practices' (Bou Akar 2018) as well as intense wildfires that have 'wiped-out 1,200 hectares of forest' (Azhari 2019).

Less than two months later, a rainstorm paralysed most of Beirut and other parts of the country, turning streets into small rivers and damaging highly impoverished working-class neighbourhoods. The severity of the flooding could have been predicted and prevented if the Lebanese government had invested in initiatives to

---

[*] The Lebanese government is constituted of a group of sectarian political parties, most of them with roots in militias. This sectarian power-sharing arrangement is the legacy of the French colonial empire that created 'a modern Lebanon' in 1920, by separating Greater Lebanon from Syria (i.e. its natural hinterland that was rich in agriculture and industry) (Tarābulusī 2012).

stop land degradation by using methods such as laying down straw mulch.* For example, the Syrian refugee camps in Lebanon's Beqaa district suffer from severe flooding every year. Having fled the war in Syria to seek safe shelter, the displaced Syrian communities not only have to endure Lebanon's harsh policing strategies† – which are used to create unwelcoming spaces and to pressure refugees into leaving the country (Sanyal 2018) – but also the worst of the effects of climate change that continue to amplify strains on failed infrastructure.

The effects of climate change in the region became apparent as early as 1999, when water shortages and desertification began to impact rural Syria. Syria was once part of the Fertile Crescent region, which possessed more than 'six million hectares of productive arable land' and was the origin of many of the world's major crops, such as wheat and barley (Hinnebusch *et al.* 2011; Pala Ryan *et al.* 2004). Now, though, the country imports an average of 1.5 million tonnes of these basic food products per year (Makieh 2018; Riabukha 2020). After a decade of intense drought, the country started witnessing crop failures, and therefore an increase in food insecurity. In 2011, 2–3 million rural Syrians were pushed into extreme poverty, including 1.3–1.5 million who were forced to migrate from their homes into urban areas (Environmental Justice Foundation 2017, p. 34). During that same year, the Syrian conflict broke out, which caused 'about half of all war casualties around the world' in the following year (Dupuy and Rustad 2018).

In this context, climate change played the role of a 'threat multiplier' (Environmental Justice Foundation 2017), as it exacerbated the existing social and political unrest, and it pushed the already vulnerable into (deeper) poverty. These heightened vulnerabilities directly resulted in an increase in child marriage cases among the displaced Syrian communities. During the Syrian conflict, child

---

*Protecting and stabilizing the soil is essential, as it allows trees to live and it reduces the risk of flooding.

† Examples of these tactics include forcing the displaced Syrian communities to 'dismantle their own shelters' concrete walls and roofs and replace them with less protective materials, or face army demolition of their homes' (Human Rights Watch 2019).

marriage became a survival mechanism among displaced families: a way to ease their economic burden by having one fewer mouth to feed (El Zein 2020) and to protect girls from the high risk of sexual harassment in informal refugee settlements (Halldorsson 2017). Reliefweb (2019) states that '41% of young displaced Syrian women are married before 18'. This illustrates how the effects of climate change are closely linked to the outbreak of conflicts, the exacerbation of existing gender inequalities and increases in poverty – food insecurity being either the direct effect or the cause among those linkages.

Lebanon is also confronted with a rise in food insecurity. Being a country with limited land and water resources, Lebanon has never been able to meet its food demands through local production alone. In addition, Lebanon currently hosts the highest number of refugees per capita in the world due to conflicts in neighbouring Palestine and Syria, thereby increasing the gap between local food supply and demand. In recent history, the country has therefore had to rely, to a great extent, on food imports to feed its ever-growing population. Due to Lebanon's recent economic meltdown, its reliance on food imports has become unsustainable. By early 2021, Lebanon's freefalling currency lost more than 85% of its value against the US dollar (Newsom 2021) compared with pre-crisis levels, while 'powerful importers and traders' continue to increase prices 'on both local and imported food through cartel behavior' (Wood *et al.* 2020, p. 2).

Furthermore, Lebanon's food supply chain is complicated by recent events: the Covid-19 pandemic, which caused import bans and quotas; and, prominently, the Port of Beirut explosion on 4 August 2020. The explosion 'caused over $5 billion in damage, killed over two hundred people, injured thousands, and forced 300,000 people onto the streets' (Peritz 2020). This deadly incident also damaged '85 percent of Lebanon's cereals', leaving 'the country with less than a month's worth of grain reserves', and it 'exacerbated one of Lebanon's core calamities: the lack of stable electricity'. While Lebanon has been suffering from power cuts for several decades, daily blackouts have increased since the incident.

By 2021, they were exceeding twenty hours a day, leaving many families – especially those who cannot afford to buy electricity from illegal generator operators – unable to use their refrigerators to store food in their homes.

According to a press release from the United Nations (2021), many displaced Syrian communities (91%) and Lebanese families (55%) are living below the poverty line. As families' vulnerability worsens and people strive to meet their basic needs, tensions between and within communities are rising due to competition over scarce resources and services (United Nations Lebanon 2021). Consequently, anti-refugee rhetoric is heightened, and this may eventually 'spark violence given the fragility of the country' (Newsom 2021).

A local media campaign called *Zarri'et Qalbi* (plant of my heart) addressed the issue of food insecurity in Lebanon by calling for the return to family farming.* Other policy reports have claimed that Lebanon's only solution is to grow 'more nutritious staples such as beans, lentils, and chickpeas, which have long been native to the region' (Wood *et al.* 2020, p. 2). Nonetheless, with no seed-producing companies present in Lebanon, local farmers will continue to rely on international seed imports and thus remain dependent on foreign-currency purchases. Furthermore, with more water shortages likely, a high risk of flooding and wildfires, as well as limited arable land, it is not enough to urge Lebanon to simply invest in local sustainable organic farming. Without the country overthrowing the entrenched sectarian system that continues to stand in the way of positive change, growing its own food and feeding its own population become increasingly unrealistic goals and more of a marketing stunt to 'piggyback' on dominant western narratives about sustainable environmental cooperation.

On a similar note, *Climate Home News,* a London-based news outlet, published an article on 17 March 2021 announcing that Lebanon aims to 'generate 18% of its electricity and 11% of its heating

---

* 'Zarriiet Albe', the Agriculture Initiative, by Nadine Labaki (www.youtube.com/watch?v=EP5fHP_Z3lM).

from renewable sources, up from a previous combined target of 15%' by 2030 (Gerretsen 2021). The article's headline was: 'While battling political upheaval, coronavirus and its worst economic crisis in 30 years, Lebanon has strengthened its 2030 emissions target'. This article is a reflection of how dominant western narratives about the climate crisis fail to recognize that the fight against climate change – and its immediate humanitarian effects – is fundamentally part of 'conflict prevention and human rights protection strategies' (Environmental Justice Foundation 2017).

At this point in time, it is no longer sufficient to narrowly focus on lowering carbon emissions in order to tackle climate change. It is also imperative to address the effects of climate change that have already materialized, especially in vulnerable regions such as Lebanon and Syria. Climate change, as a threat multiplier, endangers not only 'the stability of ecological systems which have sustained human life for thousands of years' but also 'the rights of those living today': social injustice is a central part of the climate crisis and it must therefore be addressed accordingly.

Bibliography

Azhari, T. 2019. Lebanon wildfires: hellish scenes in mountains south of Beirut. *Aljazeera*, 16 October (https://bit.ly/3s8AEUU).

Bou Akar, H. 2018. *For the War Yet to Come: Planning Beirut's Frontiers*. Stanford University Press.

Dupuy, K., and Rustad, S. A. 2018. Trends in armed conflict, 1946–2017. *Conflict Trends*, 5. Oslo: Peace Research Institute Oslo (PRIO).

El Zein, R. 2020. Rethinking the relationship between child marriage and failed infrastructure during the Syrian conflict: a discourse analysis of Arab television news. In *geschlecht_transkulturell*, edited by Hausbacher, E., Herbst, L., Ostwald, J., and Thiele, M., pp. 283–301. Springer.

Environmental Justice Foundation. 2017. Beyond borders: our changing climate – its role in conflict and displacement. Report (https://bit.ly/3JC7PpU).

Fregonese, S. 2019. Rage against the sectarian machine. *Urban Violence Research Network (UVRN)*, 29 October (https://bit.ly/3502SbJ).

Gerretsen, I. 2021. Lebanon increases climate goal despite political and economic turmoil. *Climate Home News*, 17 March (https://bit.ly/33ZczXB).

Halldorsson, H. 2017. Syrian children forced to quit school, marry early to survive. Website Article, UNICEF, 23 January (https://uni.cf/3v8acfX).

Hinnebusch, R. A., El Hindi, A., Khaddam, M., and Ababsa, M. 2011. *Agriculture and Reform in Syria*. University of St Andrews Centre for Syrian Studies.

Human Rights Watch. 2019. Lebanon: Syrian refugee shelters demolished. Report, 5 July (https://bit.ly/3JPKRvD).

Makieh, K. 2018. Exclusive: Syria to import 1.5 million tons wheat – minister. *Reuters*, 25 June (https://reut.rs/3p9637I).

Newsom, N. 2021. Aid millions wasted in Lebanese currency collapse. *CashEssentials*, 24 March. (https://bit.ly/3sVkygJ).

Pala, M., Ryan, J., Mazid, A., Abdallah, O., and Nachit, M. 2004. Wheat farming in Syria: an approach to economic transformation and sustainability. *Renewable Agriculture and Food Systems* 19(1), 30–34 (doi: 10.1079/RAFS200356).

Peritz, A. 2020. Beirut's port explosion reveals underlying problems in energy supply. *Atlantic Council*, 10 August (https://bit.ly/3BG6Nq0).

Reliefweb. 2019. UNHCR Jordan factsheet – February 2019. Factsheet, UN High Commissioner for Refugees (https://bit.ly/35jZM2c).

Riabukha, A. 2020. Syria facing wheat supply shortfall despite 27% production rise. Website, 23 December, Fastmarkets Agricensus (https://bit.ly/3KXh5qc).

Sanyal, R. 2018. Managing through ad hoc measures: Syrian refugees and the politics of waiting in Lebanon. *Political Geography* 66, 67–75 (doi: 10.1016/j.polgeo.2018.08.015).

Tarābulusi, F. 2012. From mandate to independence (1920–1943). In *A History of Modern Lebanon*, pp. 88–109. London: Pluto Press.

United Nations Lebanon. 2021. Crisis response plan appeals for $2.75 billion to respond to the impact of the Syrian crisis in Lebanon. Press Release, 12 March, United Nations Lebanon (https://bit.ly/3NkItzm).

Wood, D., Boswall, J., and Halabi, S. 2020. Going hungry: the empty plates and pockets of Lebanon. *Triangle*, 12 May (https://bit.ly/3H8vzQD).

GLOBAL SOUTH PERSPECTIVES
By Lyda Fernanda Forero

Just transition has become an increasingly well-known and accepted term as part of the necessary response to the climate crisis. From its inclusion in the preamble of the Paris Agreement and the following development in different COPs from the UNFCCC, the debate around the concept of just transition is increasingly complex, with different perspectives ranging from those of transnational corporations to ones from inviduals and the working classes.

The just transition concept originated in the 1970s and came from interaction between US grassroots community organizations and unions as they formulated proposals for communities and workers facing the closure of a nuclear power plant. The concept then continued to develop and be enriched by exchanges between organizations and social movements, eventually becoming the primary demand of the trade union movement in the negotiations on climate change (at COP15 in 2009, for example).

It was trade union advocacy that made the main contribution to its inclusion in the Paris Agreement preamble. However, just transition is not a static concept. For example, it is worth mentioning the vision of the Trade Union Confederation of the Americas (TUCA) in exchange with allied organizations and social movements:

> For the trade union movement the starting point of a transition from a high-carbon economy, based on agribusiness and mining-energy extractivism, to a socially and environmentally sustainable one ... involves ensuring that its outcome is the strengthening and expansion of decent work.
>
> CREAT (2018)

We must acknowledge the disagreements surrounding the concept of just transition, as well as the principles, dimensions and aspects that were built from social and popular movements. These emerged from processes of exchange and debate around the responses to the climate crisis from an economic, social and environmental justice perspective.

First, it is necessary to recognize the asymmetries and inequalities that are inherent in the dominant economic, political and social system at the global level. There are no single, uniform answers that work in all countries. A transition proposal that ignores the social and environmental injustice that has contributed to generating this crisis can only deepen it. For example, the reduction of energy consumption cannot be demanded with equal intensity in countries where access to energy for the satisfaction of basic needs is not guaranteed for the entire population. In order for the transition to be fair, it cannot deepen existing inequalities.

Hence, just transition is mainly a class issue in a debate about the energy system:

> Just transition aims to focus the conversation on the question of power. In other words, who controls, and who benefits from, society's use of its resources in relation to the energy system. Viewing the conversation from this angle means recognising that the current system concentrates power and the benefits of resource use with relatively few actors, while distributing the costs (including environmental destruction) across the majority, but with more impact on those who are more marginalised. This opens the door for a broader discussion of what both resource use and work ought to look like in a just society.
>
> Transnational Institute (2020)

From the perspective of the working class, a just transition of the energy system must be based on justice, democracy and people's sovereignty.

The current system is based on different forms of exploitation, discrimination and dispossession. These forms of oppression most affect the working class, and they are compounded when other dimensions of discrimination – related to gender, sexuality, race, ethnicity and age, for example – enter the mix. Moreover, the appropriation of labour and the dispossession and exploitation of nature (which is assumed to be an infinite resource) have served as the basis for the development of the current system. Just transition, in contrast, is feminist and anti-racist. It requires us to rethink the relationship between society and nature, with recognition of the need for a harmonious relationship based on principles of sustainability and of social and environmental justice, where humans are part of nature. We must also recognize in this transformation that work is an essential feature of human fulfilment.

From a working class perspective, people's movements and organizations from the Americas have contributed to defining some elements of the just transition proposals. These proposals promote decent work; guarantee respect for human rights; consider the different needs of the countryside and of urban areas; promote the democratization, de-concentration and de-commodification of energy; and guarantee free and universal access to public services, recognizing that energy is a right. The proposals emerge from democratic mechanisms involving the working class, and they acknowledge the common but differentiated responsibilities in the cause of the climate crisis. They also recognize the need to build peace in the territories; therefore, they demand the right to land and food, and they promote both agrarian reform and the overthrow of patriarchal social structures.

Democracy is a precondition for just transition. Beyond fulfilling the minimum formal aspects of a representative democracy, a guarantee of human rights, both civil and political, as well as economic, social and cultural rights enables people to participate and decide on the course of change in societies. In many places, the exercise of democracy based on the minimum requirements of defending life and opposing exploitation projects does not exist – instead we see authoritarian and fascist responses that deepen exploitation, exclusion and, with them, the climate crisis itself. Democracy should imply 'the many' questioning the control of 'the few' over production, technology, nature and energy.

In the necessary transformation of the energy system, just transition is not limited to a shift from fossil fuel to so-called renewable energy sources. While this change is necessary, it is not enough, because the ownership, use and control of the system must also be evaluated. The historical questions of energy for what and for whom must guide the just transition process, which should respond to the interests of the people, particularly the working class. The generation, use and control of energy must be public, based on democratic, sustainable and distributive principles. It is not possible to sustain a transition based on corporate interests, which have dominated the current energy system and are the root cause of the environmental, climate and social crisis.

One of the pillars of the mainstream energy, economic and social system is the trade and investment regime, which is under the control of transnational corporations. The amount of goods transported for international trade is responsible for a large part of the world's energy demand. Free Trade Agreements and Investment Protection Treaties are instruments signed between countries or blocks of countries through which the trade and investment regime acquires the binding character of an international treaty. These agreements grant wide-ranging rights to transnational corporations when making investments in other countries, protecting existing and future investments as well as profit expectations. In the case of Latin America, most of these investments are in the extractive mining, energy and agro-industrial sectors.

One of the mechanisms that ensures compliance with the rights granted to investors in such agreements is known as the Investor–State Dispute Settlement system. It allows investors to sue states before international arbitration tribunals when they feel their rights and profits have been affected (Verheecke *et al.* 2019). Many of the lawsuits filed against states relate to measures taken to curb projects in the area of energy that would have negative impacts on the territories and their communities.

Free trade and investment protection agreements thereby threaten the possibility of advancing a just transition in the following ways. Firstly, they can prevent governments from promoting policies towards just transition. This is known as 'regulatory chill'. Currently, threats of costly lawsuits against governments are considered to occur more frequently than the lawsuits themselves. For instance, they protect corporate projects for the extraction and generation of fossil fuels and minerals, both presently and in the future. If any government decides to close down projects for extracting gas, coal, oil and other minerals used in power generation, it could be held liable by foreign investors for losses under Free Trade Agreements.

Secondly, they protect subsidies to energy corporations. The extraction of fossil fuels has millions of pounds in subsidies, which take the form of tax exemptions, direct financing and concessions at prices below market prices, among others. Investment protection agreements allow corporations to sue states if they reduce existing subsidy programmes (Transnational Institute, Ecological Workshop 2020).

The dispute over the concept of just transition is, in short, a dispute about power over the energy system, which is fundamental to the economic and social system. The urgency imposed by the climate crisis requires us to question the structural causes that have generated it and, at the same time, formulate creative proposals that focus centrally on the sustainability of life.

Bibliography

CREAT. 2018. En su tercera edición, Conferencia Regional de Energía, Ambiente y Trabajo de la CSA ratifica su compromiso con el desarrollo sustentable. Report, 22 October (https://bit.ly/3x6zOuO).

Transnational Institute. 2020. Just transition: how environmental justice organisations and trade unions are coming together for social and environmental transformation. Workshop Report, February, Transnational Institute.

Verheecke, L., Eberhardt, P., Olivet, C., and Cossar-Gilbert, S. 2019. Red carpet courts: 10 stories of how the rich and powerful hijacked justice. Report, 24 June, Transnational Institute (TNI) (https://bit.ly/3uYmYvS).

PART II
European Regions on Their Way

Accelerating just transition from Southern Europe: the case of Spain

Raúl Gómez and Joaquín Nieto

THE FIRST STEPS OF THE JUST TRANSITION IN SPAIN

In recent years, the Spanish government has taken just transition very seriously and has developed an entire administrative structure around it, making it one of the best examples in the world of implementing this concept. For this reason, it is worthwhile to analyse the evolution and development of just transition policies in Spain. Later on, we will comment on the country's first successful endeavours (the closure of Spanish coal mines and coal-fired power plants) and we will talk about future prospects, but first let us look back a little to see how the current situation has come about.

Before going into detail, it must be noted that the global changes that are already underway, in terms of decarbonization and digitalization, and which we will be discussing, are being supervized by supranational institutions and organizations, rather than being left to the whim of the market or to the individual will of states. There is a difference between previous shifts in energy sources and the decarbonization process that has been undertaken globally: 'the uniqueness of decarbonisation is that it is intended to be a guided process, intentionally driven by governments and economic actors and even individuals' (Maldonado 2020). This necessary governmental action and the involvement of all actors is what gives just transition a real space in which to develop, allowing it to become binding, as well as a driver of change and a corrective to undesirable economic and social consequences.

For this reason, just transition is not a concept that, like many others that seek to combat social injustice, first matured among civil society, but one that gained importance in the heat of the Conferences of the Parties (COP) and in the United Nations Framework Convention on Climate Change, which was adopted in 1992 and which came into force two years later. From the beginning, it became clear that reducing emissions would entail considerable changes in the way almost all economic areas operated, especially in the energy industry, leading to considerable job losses. This link between energy transition and employment brought the trade unions to the forefront of the just transition field and, in the context of these COP summits, they were the first entities to grapple with the concept and to advocate the establishment of just transition policies. It is true that American trade unions were reluctant for years, but they ended up giving their support, understanding that the position of trade unionism in other countries was more sensible. In the case of Spain, where the government had not yet decided to take just transition seriously, it was the trade unions who continued to work resolutely on the concept and its implementation. In particular, the work done by the trade union Workers' Commissions (CCOO)* should be highlighted.

Thus, at COP3, where the Kyoto Protocol was adopted in 1997, Joaquín Nieto, leader of CCOO, who supported the adoption of the Protocol and a just transition approach, was also the representative of the European Trade Union Confederation. Nieto exercised leadership that confronted the position of US trade unions opposing the adoption of the Protocol. This required the development of dialogues with social partners in each country in order to anticipate the positive or negative effects of mitigation measures.

Although the Kyoto Protocol was adopted in 1997, it did not come into force until 16 February 2005, when a sufficient number of countries committed themselves by signing up to it. At that point, one of the most important practical steps was taken in Spain. When Cristina Narbona was Minister of the Environment, an agreement was reached from the initiative of trade unions that institutionalized and articulated a specific social dialogue linked to the establishment of the European emissions trading mechanisms in Spain, with the aim of achieving the commitments set out in the Kyoto Protocol. Thanks to this agreement, the decree that developed these

---

*CCOO stands for Comisiones Obreras.

mechanisms institutionalized a model of dialogue through a general round table discussion and several sectoral round table discussions, with the participation of the ministries of the environment, labour and industry. Also present were representatives from the Spanish Confederation of Employers and Industries of Spain (CEOE)* and the Spanish Confederation of Small and Medium-sized Enterprises (CEPYME),† with their respective sectoral business organizations, as well as representatives from the main trade unions (the CCOO and the General Union of Workers (UGT)‡) plus their federations. The objective of these round table discussions was to jointly analyse the employment effects of climate mitigation measures, acting preventively on possible adverse effects and proactively on the potential opportunities. This laid the foundations for the multisectoral dialogue-based mechanism that would be adapted for the just transition projects launched from 2020 onwards.

In 2007, the European trade union movement also made its contribution through a report jointly prepared by the Trade Union Institute of Work, Environment and Health of the CCOO trade union (ISTAS – Spain), Sindex (France) and the Wuppertal Institute (Germany), which was entitled 'Climate change and employment: impact on employment in the European Union-25 of climate change and CO_2 emission reduction measures by 2030' (European Trade Union Confederation 2006). Using various measurement methodologies, the report took stock of the impacts of climate change in the different European regions, based on a projection of a moderate change scenario. The European Trade Union Confederation endorsed the conclusions regarding these impacts on jobs.

The outbreak of the 2008 financial crisis altered all labour policy priorities and strategies in the face of an assault on labour rights brought about by the austerity policies enacted by the Troika (the European Commission, the European Central Bank and the

---

*CEOE stands for Confederación Española de Organizaciones Empresariales (Spanish Confederation of Business Organisations) and brings together the country's major business groups.

† CEPYME stands for Confederación Española de la Pequeña y Mediana Empresa.

‡ UGT stands for Unión General de Trabajadores (General Union of Workers). CCOO and UGT are the unions with the largest presence in Spain, and they also have the largest number of affiliated workers.

International Monetary Fund). The framework was not at all favourable to further progress on just transition strategies. Despite this, though, in the context of the climate summits, trade unions managed to keep the just transition concept alive and even to develop it further, to the extent that during the drafting of the Paris Agreement (reached at COP21 in December 2015) they convinced the Parties to include just transition and the creation of decent work in the preamble to the Agreement.

JUST TRANSITION AGREEMENTS IN SPAIN

As far as Spain is concerned, 2018 was a turning point in terms of climate change commitments. The Ministry for Ecological Transition was created, with the appointment of Teresa Ribera as minister. Ribera is an internationally recognized figure for her expertise on the climate agenda, and she was one of the architects of the Paris Agreement. Her presence also fostered a strong social commitment, and she surrounded herself with a competent team of people with great knowledge and experience of the socio-labour dimension of climate change and just transition.

This commitment was evidenced by the creation of the Just Transition Institute and the election of Laura Martín Murillo as the person who was to be in charge of it, along with the just transition agenda, in the minister's team. Laura was co-founder and director of the Sustainlabour Foundation (International Labour Foundation for Sustainable Development), in which she co-organized the first global Trade Union Assembly on Labour and Environment, which was held in Nairobi in 2006, as well as the second assembly, which was held in 2012 in Rio de Janeiro on the occasion of the Rio+20 Conference. It was at the latter conference that the United Nations adopted the concept of just transition for the first time. Laura was also a co-author of 'Resolution concerning sustainable development, decent work and green jobs',* which was adopted at the International Labour Organization (ILO) Conference in 2013, from which the ILO Guidelines on just transition were subsequently developed.

---

* See https://bit.ly/3DfXUEy.

Therefore, in 2018, the ILO and the Spanish government established a partnership that has yielded fruitful results. Through communications between Minister Ribera and ILO Director-General Guy Ryder, both entities became strategic partners for applying the ILO Guidelines during the energy transition process in Spain, and also for leading international efforts and negotiations in favour of decent work and just transition, on the path towards environmentally sustainable economies and societies. This alliance is now showing important, positive and visible results in both areas of action.

In February 2019, the Spanish government presented the Strategic Framework for Energy and Climate, composed of three main instruments: a draft Climate Change Law, approved in 2021; the National Integrated Energy and Climate Plan (PNIEC[*]); and the Just Transition Strategy (Estrategia de Transición Justa[†]). Although these instruments can always be improved, they nevertheless provide a solid regulatory framework for the decarbonization of the Spanish economy.

In this context, the Spanish Just Transition Strategy aims to respond to the socio-labour impacts generated by the transition processes. This strategy, which adopts the ILO Guidelines, establishes mechanisms to promote the creation of decent work together with social cohesion, addressing vulnerable groups and the economic sectors and territories that will be most affected by the energy transition. A strategy such as the one proposed is critical in Spain, due to the high unemployment rate (double the European average), the deficits in quality employment, and the social inequality and in-work poverty, all aggravated by the financial crisis. None of these things can be neglected in energy transition policies. Furthermore, the strategy warns that the energy transition affects not only particularly vulnerable populations, but also territories and sectors that have benefitted from the fossil fuel industry. For this reason, the strategy contemplates an urgent action plan for coal regions and power plants undergoing closure, in order to govern the energy transition process and to address the social impacts of these policies. It should be noted that the process consists of not only a declarative or documentary process but also a set of negotiations and agreements with local governments, companies, affected social

---

[*] PNIEC stands for Plan Nacional Integrado de Energía y Clima.
[†] Available at www.transicionjusta.gob.es/common/ETJ_ENG.pdf.

partners, trade unions and employers. The government also considers actions aimed at closing mines and thermal power plants in a specific timescale while promoting effective social protection measures and new investments and economic activities that generate employment, involving energy companies in the process.

A transformation as disruptive as the energy transition requires innovative approaches, processes and mechanisms. The Spanish experience is a true example of social innovation, articulated around two innovative mechanisms: the Institute for Just Transition (ITJ*) and Just Transition Agreements. The ITJ is the Spanish Government's instrument for developing its Just Transition Strategy. It promotes and manages the energy transition processes and its social effects through social dialogue at all levels, mainly with trade union and employers' confederations and federations, the companies and workers affected and their representatives, and regional and local authorities. Not only does it proceed with the restructuring that decarbonization and the ecological transition require, i.e. the urgent closure of mines and coal-fired power stations and the dismantling of nuclear power stations, but it also promotes economic activities with sustainable and inclusive characteristics, to replace the economic activity that disappears. This generates more and better jobs, as well as a more diversified, more inclusive and more sustainable framework of production. The most innovative mechanism being implemented is the Just Transition Agreements.

The ITJ defines these Agreements as follows:

> The priority objective of the Just Transition Agreements is the maintenance and creation of activity and employment, as well as the fixation of population in rural territories or in areas with thermal or nuclear facilities in closure. To this end, they promote diversification and specialisation consistent with the socio-economic context and provide at-risk sectors and groups with tools to support investment, the restoration of territories, support for industrial projects, the retraining of workers and the development of SMEs in order to achieve their objectives.

---

* Available at www.transicionjusta.gob.es/ (the website is in Spanish but many of the documents are available in English).

Their elaboration follows participatory processes in which an objective assessment of possible job losses and a commitment is proposed, with a final list of measures and projects to maintain employment and population. In other words, to lay the foundations for a sustainable future project for these territories that were fundamental actors in the generation of today's wealth and whose contribution we must recognise, respect their identity and help them to continue to play a leading role in the economic future of our country.*

There are more than a dozen agreements spread across the Spanish autonomous territories that are most affected by the closures (Galicia, Asturias, Aragon, Andalusia, Castilla-La Mancha, and Castile and León). What is most interesting is that the number of companies and workers affected by the closures is far smaller than the number of companies and jobs contained in the thousands of replacement public and private initiatives that are already being mobilized.

JUST TRANSITION AGREEMENTS

1	JTA Garoña	8	JTA Suroccidente Asturias
2	JTA Zorita	9	JTA Puente Nuevo - Valle del Guadito
3	JTA Aragón	10	JTA Los Barrios
4	JTA As Pontes	11	JTA Carboneras
5	JTA Meirama	12	Montaña Central Leonesa - La Robla
6	JTA Valle del Nalón	13	Guardo - Velilla
7	JTA Valle del Caudal	14	Bierzo - Laciana

Figure 1. Areas in which just transition conventions have been implemented. (*Source*: https://bit.ly/33FhfS2).

---

* See www.transicionjusta.gob.es/Convenios_transicion_justa/.

The finalization of the participatory processes and the signing of these Just Transition Agreements protocols coincided with the global Covid-19 pandemic. Contrary to what might have been expected, this did not lead to a delay in implementating the agreements, and the committed budget allocations did not undergo any variation.

The table in figure 2 shows the status of these agreements as of March 2021.

Finally, with regard to the choice of sectors and regions in which to implement these agreements, it is also worth asking whether only social justice criteria have been applied, or whether these plans are-only aimed at the most contentious sectors that could cause the most uproar and thus hinder the transition to a new energy model. For the moment, the answer is simple: the Spanish government has started with the most urgent sectors. Given the need to decarbonize society, the first thing to do is to put an end to the industries that emit the most greenhouse gases per unit of energy produced, i.e. coal. It remains to be seen whether in the next steps towards a just transition we are able to keep the quest for social justice and the reduction of inequality as priority objectives, while also taking into account the arguments of the different collectives, because this is where the real causes of social tensions lie.

PERSPECTIVES

It is still too early to make a definitive assessment of the Just Transition Agreements, but, as we have already explained, they are developing more than satisfactorily so far, and we can affirm that they are setting an example for other productive sectors and countries to follow.

However, looking ahead to the next few years, there is one element that cannot be ignored in any analysis: the effects of the projects that the Spanish government may implement within the framework of the Next Generation EU funds. Apparently, matching the Just Transition Strategy with these projects should not be particularly complicated. Furthermore, in the Recovery, Transformation and Resilience Plan (RTRP), which the Spanish government presented to the European Commission to obtain the funds (and which was

Autonomous Communities	AGREEMENTS	PROTOCOLS	DIAGNOSIS			PARTICIPATORY PROCESS		
		Signature date	Preliminary diagnosis and delimitation	Final diagnosis	End date	Number of stakeholders involved	Proposals and ideas submitted**	Participation Report
Aragón	Andorra-Mining Regions	22/05/20	YES	YES	07/06/20	69	140	YES
Asturias	Southwest	25/03/20	YES	YES	13/12/19	33	78	YES
	Caudal	25/03/20	YES	YES	21/06/20	39	67	YES
	Nalón	25/03/20	YES	YES	21/06/20	37	75	YES
	Bierzo Laciana*	20/11/20				179	508	
	Bierzo Alto		YES	YES	31/07/20	54	124	YES
	Laciana–Alto Sil		YES	YES	31/07/20	45	140	YES
Castilla-León	Cubillos-Ponferrada		YES	YES	31/07/20	43	127	YES
	Fabero–Alto Sil		YES	YES	31/07/20	37	117	YES
	La Robla	20/11/20	YES	YES	31/07/20	41	147	YES
	Guardo-Velilla	20/11/20	YES	YES	31/07/20	36	184	YES
	Carboneras	09/03/21	YES	YES	06/09/20	22	49	YES
Andalucía	Puente Nuevo	09/03/21	YES	YES	31/07/20	37	221	YES
	Los Barrios	09/03/21	YES	In preparation	30/04/21			
	Meirama		YES	YES	31/07/20	28	65	YES
	As Pontes		In preparation			In preparation		
C.León / Basque Country	Garoña		YES			In preparation		
Castilla la Mancha	Zorita	24/11/20	YES		31/01/21	29	66	In preparation
					Total	550	1600	

* For the JTA of Bierzo–Laciana 4 different participation processes were launched

** Total number of all proposals received, subject to change as data is processed or new proposals are received.

Figure 2. Status of the Just Transition Agreements as of March 2021. (*Source:* Just Transition Covenants. See https://bit.ly/3IdTXSv).

applauded by the latter), just transition has a prominent role. In the RTRP it is stated that just transition is essential, given the scope of the transformations that are coming in several industrial sectors. Specifically, it argues that just transition is fundamental to five of the six pillars that underpin the entire RTRP. It will be necessary for the specific projects that are drawn up to consider how to guarantee a transversal application of just transition measures and also to consider whether the RTRP budget allocation for these (€300 million, which will be added to the European Just Transition Fund) is sufficient for all the actions that will be added to those already planned by the Just Transition Institute.

It remains for us to comment that this concept of 'energy transition', to which the EU has made a firm commitment, is intertwined with another transition: the digital transition. While until now, when we have talked about a just transition, we have done so almost exclusively in reference to energy and the obligation to opt for sources that allow us to reduce greenhouse gas emissions, we must now acknowledge that this second (and simultaneous) transition is beginning to affect many more sectors.

Any process of technological change entails a disruption of work. These processes of change have been perceived by large sections of the population as processes of 'creative destruction', to use the expression developed by the economist Joseph Schumpeter (1883–1950). According to this idea, technological change always leads to job destruction in its early stages but, over time, new technologies eventually generate jobs, and these may become more numerous than those destroyed.

This theory may, though, be blown out of the water on this occasion. The digital transition – i.e. the intensification of the application of technologies to all spheres of administration, security and life itself (the frequently mentioned 'internet of things') – could lead to a severe technological gap, not just in leisure and communications, but in all areas of life, and this could leave so many people behind that it would not be compensated for by just transition policies like those currently being planned. This might be more evident in Spain than in other countries because of the large areas of the country, far away from large population centres, that suffer from depopulation.

This situation will have to be included in the equation, to make sure that the technological gap does not increase even more in these rural areas.

We should bear in mind that the digital transition is already underway, although a large part of the population in Spain behaves as if it did not exist, tending instead to maintain production and business models that will become obsolete (if they are not already) in the face of the digital transition. This inertia will clash with the transition for two reasons: firstly, having been promoted by European institutions with a veritable outpouring of millions, it is going to spread quickly and widely; and secondly, everything linked to this technological change is developing at breakneck speed. And let us not forget that when we talk about the digital transition we are talking about practically all economic sectors.

In the area of labour market robotization, studies and predictions are being made about its impact on employment, but in the area of digitalization, we find that it is much more complicated to delimit its sphere of influence; furthermore, society's behaviour in this area is far from an exact science. So, for those of us who have the right transition between our study and work priorities, there is still a lot of hard work ahead to prepare for the expected digital tsunami.

Finally, we do not want to close this chapter without mentioning another risk that we must bear in mind: fighting climate change without rethinking economic growth strategies will potentially lead us to an undesirable scenario. We cannot elaborate on this because it is outside the scope of this analysis, but considering the physical limits of the materials required for these transitions and the environmental impact of obtaining them is an area of much-needed reflection, and one that is already underway. At the same time, it should not be forgotten that, at present, extractivism at a global level is one of the factors that maintains North–South inequalities; thinking ahead, the EU is seeking to achieve a certain autonomy in the production of these necessary raw materials, which is encouraging the emergence of hundreds of new mining projects within European borders (with several hundred of them in Spain).

Governments and institutions have finally started to fight climate change decisively, meaning that we are heading towards a serious and uncertain alteration of our economic and productive model. The objective of a just transition is not to generate more inequality in a society that is already very unequal. The Spanish government has begun to work in the right direction, but we must ask ourselves whether these efforts will not be dwarfed by the changes that lie ahead.

Bibliography

European Trade Union Confederation. 2006. Climate change and employment: impact on employment in the European Union-25 of climate change and CO_2 emission reduction measures by 2030. Report, ETUC, Brussels.

Maldonado, M. A. 2020. Transición Justa: y ahora ¿qué? *Ethic*, 25 May (https://bit.ly/3h9LNyF) (In Spanish).

Historical perspectives to inform a just transition in Southeastern Europe

Dragan Djunda, Aleksandar Gjorgjievski and
Vedran Horvat

TRANSITION IN SERBIA: SMALL HYDROPOWER PLANTS FOR LARGE PROFITS
By Dragan Djunda

Small hydropower plants (SHPPs) are based on the nominally renewable power of water but they are a highly unsustainable technology. The impossibility of a trade-off between renewability and sustainability became obvious in the region of the Western Balkans, where almost 3,000 SHPPs were planned and hundreds of them have already been built. The main issue there was that these SHPPs were installed on pristine rivers and streams in mountainous, often protected areas. These are some of the last wild rivers of Europe, rich in rare and protected species. The way in which SHPPs operate endangers these invaluable habitats as well as the local communities living near the rivers.

Based on data from fieldwork in the villages of the Stara Mountains (Serbia), we will illustrate why SHPPs are a deeply unjust energy technology and how they were successfully contested. We will argue that SHPPs were implemented as a first step in the energy transition in Serbia, but that they had a very limited transformational capacity for the energy system. Therefore, they maintained the status quo and remained merely profit-making machines for international and domestic capital. As such, the case of SHPPs reveals that the structural ecological, economic and social issues involved are unlikely to disappear with a less harmful technology.

The economic and ecological harms of SHPPs

The investments made in SHPPs negate all of the key International Labour Organization (ILO) guidelines for a just energy transition (Sabato and Fronteddu 2020). This applies to the rules concerning the distribution of economic and environmental costs and benefits as well as to the procedures for providing socio-environmental protection and ensuring wide participation in the decision-making process.

SHPPs in Serbia produced deep divisions in ecological and economic costs and benefits: they brought fear for survival to the affected communities but easy profits to investors. While the water flowing through the pipes meant profit for investors, for the locals it represented the essence of life and a threat that such a life could end. In the mountains across Serbia, pipes and turbines have left the water contaminated and many riverbeds dry (Ristić *et al.* 2018). This means that numerous protected species in the Stara Mountains – such as stream trout, river crabs and otters – could be endangered as well. It was almost universal that SHPPs were built with significant flaws such as dysfunctional fish passages and pipes built into the riverbeds. The reason for these flaws could have been a lack of expertise, but it is difficult to ignore the fact that these errors increased the amount of water in the turbines, thereby producing more electricity and bringing more profit to investors.

The communities in the Stara Mountains were concerned that the quality of drinking water would be affected too. They feared that some streams would dry out due to the imbalance created in underground waters, causing further issues such as erosion. Citizens argued that the SHPPs would also affect their food production because many people either used water from the rivers for irrigating their gardens or thought that the rivers and underground waters were providing the right levels of moisture for their land. As the citizens had small pensions, the vegetables and fruits they grew were important for their survival, either for personal consumption or small-scale sale. Moreover, communities in these mountains had a long history of cattle breeding, for which the abundant freshwater streams were essential. With contaminated and reduced levels of water, sheep, cows and goats would have much more limited access to clean and abundant water.

The economic subsistence of the villages further suffered as dams and pipes destroyed the touristic potential of the area. Tourists, hikers and fishermen often visit the Stara Mountains for their natural beauty. They rent short-term accommodation in the villages, providing additional income to the local population. Many people in the area have renovated their family houses to accommodate visitors, and they were understandably worried about the future of rural tourism if the numerous waterfalls, springs and intact rivers of the region were spoiled by SHPPs.

The fear of depopulation existed even before the SHPPs. Like other rural post-socialist regions, the Stara Mountains were facing the disappearance of their population due to decades of rural–urban migration (Dzenovska 2020), and the trend has intensified since the 1990s, when the possibility of profitable agricultural production vanished. The decaying infrastructure and the vacated buildings and homes were the best evidence of this rural emptiness. The investors and institutions that approved SHPPs exploited this context of exodus, anticipating that the elderly population would not be able to resist their plans. Simultaneously, this context allowed investors to falsely promise that the SHPPs would bring a revival to the villages through employment, renovated schools and roads. The case of SHPPs initiating the energy transition therefore demonstrated how, instead of improving conditions in peripheral regions, they only deepened the region's underdevelopment.

Activists from the Let's Defend the Stara Mountains movement, from non-governmental organizations (NGOs), from the academic community, and concerned citizens articulated the problem with SHPPs and demanded cancellation of all projects in the country. Unfortunately, they achieved only the termination of plants in the Stara Mountains, where the protests were strongest. Besides protests, the main strategy used against the institutions and investors was promotion of an alternative vision of sustainable rural development. If the SHPPs brought further exploitation, this new vision materialized in other ways: the numerous revitalizations of cultural and educational institutions in the villages, renovations of homes, and calls for the repopulation of the mountains.

Procedural and recognitional justice

None of these ecological and socio-economic harms would have been so extensive had the procedures that were supposed to protect the environment and its communities operated properly. These procedures were nominally introduced to ensure that the environment was not affected by SHPPs, which was especially important since many plants were planned in protected zones. Additionally, they required investors to consult the local communities and get their approvals for investments.

In reality, these protective procedures were illusions. The residents in the Stara Mountains complained that they were not substantially involved in the decision-making. The local authorities and investors often skipped public consultations with them or invited only a handful of favoured residents. Even when consultations did occur, the authorities did not provide a full account of the design and consequences of SHPPs. Instead, they misrepresented the plants as watermills and promised benefits to the villages while hiding the technical and ecological details of the plants' operation.

Similarly, environmental protection was only provisionary. For example, SHPPs below 1 megawatt were allowed to be built without an environmental study, and some of the studies were being conducted by biased professionals who were close to the investors. There was therefore a discrepancy between what was written in the studies and the real situation on the ground; for example, they often falsely negated the existence of protected species in the areas. Moreover, activities that were against the environmental protection rules were conducted, such as building pipes in riverbeds, having dysfunctional fish paths or destroying the surrounding forests.

It is now evident that no formal mechanism of control operated properly. The reasons for this might be weak institutions, a lack of coordination between the institutions involved, or the strength of the actors from the energy sector. The laws and procedures represented only formal confirmations of the legality of projects for national and local authorities. They were also justifications for investors and proofs of sustainability for international banks and financial institutions, which poured credit into these projects.

Even beyond these protective procedures, the planning for SHPPs points to a deep division in knowledge, values and interests that the current energy transition is built upon. The principles of profit, technological efficiency and the maximized use of resources oppose the values of sustainable development, biodiversity and the culture and economy of rural life that is dependent on rivers. The stream of engineers that prioritized energy as a resource, in cooperation with the hydropower lobby, had a decisive role in determining the number of SHPPs and their design, excluding the values of local communities, biologists and sociologists. In that way, the ultimate purpose of SHPPs was profit, while the survival of species and communities was reduced to the bare 'biological minimum'. Cases such as this were often employed to demonstrate how profit-making and sustainability were compatible, and even mutually supportive.

The opponents of SHPPs often emphasized how destructive the biological minimum principle was. Some of the most recognized opponents were scientists and academics who countered the dominant claims and principles. They conducted field research on the locations of SHPPs, demonstrating that the conflicts between profit and life were unresolvable. Also, activists and NGOs organized alternative consultations with local communities in each village of the Stara Mountains, giving voice to everyone who was affected by SHPPs. Moreover, they promoted the values of biodiversity and rural development to counter the dominant technoscientific principles. Importantly, the opposition to SHPPs achieved something far beyond simply defending against SHPPs being built. Through their activities, ecological activists, scientists and citizens brought ecological issues to the forefront of Serbian politics. They framed the ecological agenda as an essential public good that exceeds political particularities and social divisions.

Profit and the status quo

The financial scheme behind SHPPs reveals why they mushroomed across the whole region.

Firstly, they required less investment than other technologies. Investments in hydropower are less capital-intensive than those in large solar and wind farms. They were also less dependent on knowledge from abroad because of the expertise developed when Serbia was part of the Socialist Federal Republic of Yugoslavia.

Secondly, as in other countries, the Serbian government introduced feed-in tariffs (FITs), guaranteeing a beneficial price over 12 years to all renewable energy source (RES) producers, including SHPP investors. It is now clear that some of these investors were close to the Serbian ruling party. SHPPs were a profitable opportunity for foreign capital, too. Most of the credits for SHPPs in Serbia came from commercial banks such as Erste Bank, UniCredit, Banka Intesa and Société Général (Bankwatch Network 2019). Large financial institutions like the European Investment Bank, the European Bank for Reconstruction and Development, and the World Bank's International Finance Corporation, together with Norwegian, Austrian, German and Italian development banks, poured hundreds of millions of euros into greenfield hydro projects in the region.

And thirdly, SHPPs were a good opportunity for the producers of hydro equipment to reanimate their industry, which was drowning in the global rush for wind and solar sources.

Although they are profitable businesses, SHPPs are an insignificant contributor to the national energy system, which is dependent on coal. If all 850 proposed plants had been built in Serbia, they would have contributed only 2–3% of the country's electricity needs (Ristić *et al.* 2018). Until April 2021, the Serbian government kept technology caps on wind and solar power eligible for FITs. This means that, with its minor capacity, SHPPs could not affect contributions to the national energy sector.

The new RES law, implemented in April 2021, was presented as a turning point by the government. It removed technology caps on wind and solar (which had previously limited how much energy was subsidized per type of technology), boosting investment in these two energy sources, while at the same time allowing citizens to participate more in the energy transition. Investments in SHPPs slowed down, but SHPPs were not banned by the new law, except for those in the Stara Mountains, where the local authorities succumbed to public pressure. However, even if new investments in SHPPs disappeared, there is a danger that their devastating social, economic and ecological consequences outlined above would remain the basis for the Serbian energy transition.

Early signs since the introduction of the new law have not been encouraging. Foreign energy corporations are already jockeying for position on the newly open Serbian market, citizens' participation is significantly limited, and the standards and procedures of protection (which broke under the pressure of the SHPPs) remain in ruins. The only hope lies in increasing ecological awareness and campaigning against SHPPs in the newly established activist network.

Bibliography

Bankwatch Network. 2019. Western Balkans hydropower: who pays, who profits? Report, September, Bankwatch Network (https://bit.ly/ 3BSmqdX).

Dzenovska, D. 2020. Emptiness: capitalism without people in the Latvian countryside. *American Ethnologist* 47(1), 10–26 (doi: 10.1111/amet .12867).

Ristić, R., Malušević, I., Polovina, S., Milčanović, V., and Radić, B. 2018. Male hidroelektrane derivacionog tipa: beznačajna energetska korist i nemerljiva ekološka šteta. *Vodoprivreda* 50(294–296), 311–317. (In Croatian.)

Sabato, S., and Fronteddu, B. 2020. A socially just transition through the European Green Deal? Working Paper, August, European Trade Union Institute, Brussels.

ENERGY TRANSITION IN NORTH MACEDONIA
By Aleksandar Gjorgjievski

The Southeastern European country of North Macedonia is also struggling with a wave of SHPP investments as a result of a resurrected, decades-long plan for building hundreds of them on its territory, including in protected areas and national parks. SHPPs are also part of the FIT system for renewable energy, which requires vast financial stimulus despite contibuting in only a minor way to electricity production. The country gets a bigger portion of electricity production from its large hydropower plants that were built decades ago, accounting for 40–50% of North Macedonia's total electricity production; the rest is mostly coal-based power plant production.

According to a report by the Health and Environmental Alliance in 2019, the sixteen coal-based power plants in Southeastern Europe produce more air pollution than those in the rest of Europe combined due to the old technology of these plants. As well as causing more air-pollution-related illness and death in the Balkans, the problem spreads across borders into most EU countries and has a huge negative impact on their health systems.

Throughout Southeastern Europe, the energy sector is mostly coal based, divided approximately fifty–fifty between coal and hydro, with a very small portion of RES. The picture is no different in North Macedonia, which in 2017 emitted 30% more CO_2 per capita than the EU average. The country's current government has set promising milestones on the energy agenda – it plans to shut down some coal-based power plants by 2025 and to completely shut down all of these plants by 2040. At the same time, by 2023 there will be a complete modernization of the remaining coal-based power plants to meet the standards for desulphurization and decarbonization in coal burning. The first steps in this transition have already been made: a solar power plant has been built on a closed coal mine in Oslomej in the western part of the country, with a planned future capacity of 100 megawatts, and the capacity of the existing state-owned wind power park in Bogdanci in the south-east of the country has been expanded. The government's strategy also involves investment in solar power plants on all closed and existing coal mines and coal-based power plants.

According to experts, the rest of the strategy is not so sustainable. With the energy sector (mostly state-owned) using the premise of having to meet basic energy needs, priority will be given to gasification (which is also based on a fossil fuel): there will be huge investment in a completely new national gasification network for most electricity production as a transition from coal to gas, particularly targeting urban areas. From the current standpoint, the country will therefore miss the opportunity to really modernize its energy sector towards long-term sustainability through the use of renewables. It will therefore be difficult to achieve a just transition focused on green employment and the energy independence of the economy.

One of North Macedonia's main ongoing challenges in terms of decommissioning lignite-burning plants is the central importance of the coal-based economy to regional and national revenue streams. State-owned coal plants provide a major source of employment for the country. They also support various ancillary industries around the extraction, processing, transport and distribution of energy resources, in addition to the development and maintenance of energy infrastructure.

So while the government has nominally announced a phasing out of coal, as discussed above, the practicalities of this transition on the ground are unclear. Among the potentially affected communities, there is very little awareness or acceptance of the major economic, political and social trade-offs that are required in order to wind down lignite-based infrastructure. What is more, the operation of two active power plants in the country (Bitola and Oslomej) is closely connected with various vested interests on a national and a regional scale. In part, this is due to the predominance of state ownership in the energy production sector, overemployment in state-owned electricity-generating industries, and the historically strong role of energy lobbies in influencing the economic policy of the country.

NGOs working in North Macedonia have been actively attempting to promote alternative sources of income and employment for coal-based economies and to start a public debate on the practicalities of the coal phase-out process. The Eko Svest NGO in particular has been running a project on just transition in the Kichevo and Bitola regions, working on the development of new forms of tourism, agriculture and local enterprise as innovative forms of employment, to create opportunities for workers in the coal industry. One of their key arguments is that a just transition away from coal does not need to involve the energy sector per se. While renewable energy installations (some of which are noticeably starting to be developed, particularly in the Kichevo region) typically lead to lower direct levels of employment, the skills and knowledge that are embedded in the coal industry can be utilized for the development of other economic sectors. Many of these other sectors (particularly agriculture) were present in the affected regions before the development of fossil fuel resources.

Active organizations in the energy sector argue that the Kichevo and Bitola regions need to see an immediate commencement of efforts to promote the long-term planning of fair transition programmes. The process necessitates the active participation of workers, trade unions, local people, small and traditional businesses, civil society organizations and local authorities. NGO actors argue that the government has a duty to support its citizens living in the transitioning coal regions as well as a duty to mobilize financial resources in order to fulfil the goals that accompany a future fair transition plan. There is an expectation that authorities should be transparent with all stakeholders. The process needs to involve private businesses and local people interested in generating renewable electricity (particularly by installing photovoltaics on the roofs of households, businesses and public buildings). Building local trust is key.

Although North Macedonia has very high solar radiation compared with its neighbouring countries, and there is therefore big potential for renewable electricity production, the country is not using this comparative advantage, and it currently has very low production levels from this energy source. Even with the promising FITs for renewable production, investment in renewables for the energy sector still remains a privilege of big businesses only, and individual energy production by citizens (households and buildings) is kept at the lowest level and only for their basic needs, with no opportunity to contribute to the country's energy sector. The country's future plans are therefore failing to follow EU trends, whose 'prosumer' approach for households creates the optimal potential for production. State regulations are instead focused on stimulating big capital only, therefore locking in the structure of the energy sector for the long term.

Moreover, establishing energy cooperatives is still very limiting, with very few possibilities for development in this direction. Only one start-up, Green Cooperative (Zelena Zadruga), was established in 2020 as a pioneer in the research and development of energy co-ops but with no investment in production. In light of this approach, the just transition will not be so 'just' from a social and economic point of view in the medium to long term because it again makes the employment market and the energy market dependent on a handful of big companies that are usually owned by international investors.

A positive step for the region's energy sector came with a November 2020 EU initiative announced in Sofia that offered post-pandemic support to the region (European Commission 2020):

> The Western Balkan countries signed the implementation of the Green Agenda for the Balkans focusing on five main areas: Decarbonization, Circular economy, Depollution, Sustainable food systems and rural areas, Biodiversity.
>
> The Green Agenda for the Western Balkans is a new growth strategy for the region, leaping from a traditional economic model to a sustainable economy, in line with the European Green Deal. It is embedded in the Economic and Investment Plan, which has a truly transformative potential and aims to spur the long-term recovery of the Western Balkans and their economic convergence with the EU. The plan will be backed by a twin green and digital transition for the Western Balkans.
>
> It sets out concrete recommendations to: Align the region with the EU's 2050 ambition to make Europe a carbon neutral continent, unlock the potential of the circular economy, fight pollution of air, water and soil, promote sustainable methods of food production and supply, and exploit the huge tourism potential of the region, focusing on biodiversity protection and restoration of eco-systems. The EU will support financially the implementation of the ambitious Green Agenda through the Instrument for Pre-Accession (IPA III*). The Western Balkans Investment Framework, the European Fund for Sustainable Development Plus (EFSD+), and other instruments will be the main implementing mechanisms in this regard.

---

*IPA III funding is subject to the adoption of the next Multiannual Financial Framework 2021–27.

Due to the huge impact of the pandemic on the fragile economies of the Western Balkans, the Green Agenda is a very promising development and is setting a pathway in a positive direction. We will, though, need some time to see how this EU support unfolds. Concerning the energy sector, the region's governments should use this chance to make their economies more resilient and less energy dependent. By supporting the prosumer model and encouraging the development of social entrepreneurship in the energy sector, they will both promote equality in their societies and preserve the environment.

On the other hand, the point of just transition and the Green Agenda is not just to undertake a simple transition from dirty energy sources to clean ones and then to lock down the energy sector with a few large investors that monopolize the energy production of the entire region; instead, the 'just' aspect of the energy transition should be done in the best way to support social and economic equality and to provide sustainability and resilience to local communities so that they can thrive and progress.

Bibliography

European Commision. 2020. Guidelines for the implementation of the green agenda for the Western Balkans. Working Paper, European Commission, Brussels (https://bit.ly/3pi5rwS)

Health and Environmental Alliance. 2019. Chronic coal pollution. Report, 19 February, HEAL, Brussels (https://bit.ly/356s6oM).

TRANSITION POSTPONED: WHEN CLIMATE URGENCY IS NOT ENOUGH FOR JUST TRANSITION IN CROATIA
By Vedran Horvat

If readers carefully scrutinize the main national strategic and policy documents relating to climate change in Croatia, they might conclude that the climate crisis does not exist and that there is no climate urgency. Despite the fact that domestic policymakers have in recent years started to recognize that collective climate action is needed to mitigate greenhouse gas emissions – evidently under pressure from

Brussels – and that the necessary actions have to be integrated into national policies, they have still failed to recognize the emergency – that the time to act is, if not yesterday, then now, today, and certainly not tomorrow.

But the main strategic documents delay any ambitious action until 2030 and ignore the need to assess the vulnerability of different sectors, such as mobility and transport, to climate change (Croatian Parliament 2020). Fortunately, the government recognizes the vulnerability of tourism, biodiversity, water, energy, fisheries and agriculture, but again they completely ignore whole sectors of the current economic production and supply system (this seems to be entirely missing from policy documents), and so they accordingly treat thousands of jobs, very often jobs with a high carbon footprint, as irrelevant. This sort of deep, structural denial stops us from entering into proper debates about the scope of just transition in Croatia. However, in the following section I will dive deeper into the context and conditions that are needed for just transition to be considered as a course of collective action.

Transition as a never-ending story

The government's denial of the climate emergency highlights concerns that Croatia is a place where only more passive and consumption-oriented modes remain relevant, and where production and human labour, including decent working conditions, are left in survival mode. As well as this, in deeply de-industrialized Croatia, where most industries have been plundered (due to a cycle of privatization justified by the ruling regime in the 1990s), there is a dispute over which industries supposedly survived the era of 'transition'.*

---

* In this corner of Europe, 'transition' is usually understood as a complex process of transformation to a market economy, the rule of law and democracy – this started with the dissolution of Yugoslavia and the war in the region in the early 1990s. It was marked with non-transparent privatization of national companies and firms, a growing arms trade and an abundance of corruption scandals. Transition was supposed to deliver a better quality of life, but instead it brought to the average citizen many doubts and concerns related to how superficial and shallow systemic change was in the name of public interest.

Political scholars such as Dejan Jović claim that Croatia has to finish a fivefold transition, and most of that work has not yet been completed. On the other hand, Jović (2021) claims that Croatia has to abandon its dual setting of a post-Yugoslav future along with TINA ('There Is No Alternative'), and he argues that new generations need to 'begin imagining and conceptualizing a new [alternative]'. Let us consider the fact that socio-ecological transformation that integrates some of the principles of just transition could be one of the avenues in which new generations could take such an initiative.

The burdens of this unfinished transition are multi-faceted, predominantly because the widespread opinion is that transition did not fulfil any of the governments' promises (Džihić and Hamilton 2012). Additionally, there is a vague feeling that this transition was never completed, because the capitalist system and democratic order entered into deep crisis in parallel with the transition process, which does not favour any new or 'other' kind of transition. This perception of lack of completion is not reduced to the language of transition as such, it is more a narrative of an unclear journey from point A to point B. There is a lack of collective understanding about the external pressures that make society develop in a certain direction. Nevertheless, the global, transnational character of climate change and planetary boundaries indeed transcend particular national trajectories. Accordingly, as climate change urgency is far from prioritized in Croatian society, there is a lack of both understanding and political will/pressure to initiate another transition: a transition to a democratic, fair and decarbonized society that is able to provide a sufficient number of decent jobs for citizens and workers.

Just (another) transition?

To the largely pauperized and de-industrialized European semi-periphery, exposed as it is to neocolonial, exploitative and extractive practices, particularly in Southeastern Europe, the notion of 'transition' is sometimes irritating. It relates to the sense of being caught in limbo, in a sort of desert or swamp, from which mass emigration is the only solution, and as such, a threat to the successful completion of transition, because these countries experience brain drain (Horvat 2004).

First and foremost, this term evokes memories of the failures of the 30-year-long 'transition' to a market economy and democracy, a period that witnessed many things: severe financial crises, austerity measures, the downfall of European democratic development with the ascent of the far right, the return of authoritarianism and further shrinkage of the democratic space, all now accelereted in the pandemic period. This transition was supposed to bring back the 'golden days' of prosperity and peace, but instead it brought the opposite.

Secondly, in this part of Europe, where climate change is not taken seriously enough by decision makers, where coal and nuclear phase outs are constantly postponed and where the overall industrial sector has been devastated, the notion of a just transition does not resonate with the idea of hope. This has to change if we are to take just transition seriously (Perspectives 2021).

Transition is therefore still sometimes perceived as another sacrifice and is marked by a collective feeling of injustice. People have tended to draw the conclusion that after a long period of war, poverty and plunder, and now that they have finally reached conditions in which to enjoy peace and well-being, they are being forced to make new sacrifices despite not feeling responsible for creating the climate crisis.

When is it enough?

Furthermore, the just transition narrative is landing in a society that is completely hypnotized by the idea of growth – indefinite growth that should bring us closer to well-being and prosperity. And as success in a neoliberal society is often identified as wealth and material abundance, it is not always easy to draw the boundaries of sufficiency, or to say that enough is enough! In many cases, trade unions are not eager to be spokepeople for such attitudes.

Additionally, many of the few industries that have managed to survive in the country are fossil-fuel-intense ones. On the other hand, industries and sectors in which a major shift toward decarbonization could be made (e.g. modernizing the railways) are often at risk of privatization or are 'captured' by clientelistic or political party networks (Perspectives 2021). Under such conditions, workers are not exactly fascinated by new buzzwords – instead they have doubts and reservations: they cannot find guarantees that their jobs and their dignity will be protected. They are gradually recognizing the potentially hazardous impacts of climate change, but merely as risks to the future of their work and their job security. Accordingly, just transition in Croatia is facing tremendous challenges: implementing policy shifts that will respond to climate change and at the same time keep industry and decent work conditions alive, or even improve them. Decision makers are not yet ready to demonstrate the political will needed to go in that direction: they are locked in a sort of denial, waiting for the worst to take place, so that they can have the justification they need to act. They are obviously not able to induce an incremental, gradual transition for most of the remaining carbon-intensive sectors and industries.*

Glimmers of hope

So where can we find hope? Where will the impetus for just transition come from? For the moment, it will come from emerging cooperatives and from cities where green political platforms have become influential. In that sense, many eyes are directed toward the capital Zagreb, where the green left political platform Možemo has gained power and has taken over the city's government (Euronews 2021). They are now faced with the challenge of a deep systemic transformation of municipal public service companies, which employ thousands of people, in order to decarbonize the local economy through ecological modernization while leaving no one behind. There are a few other cities, such as Pula, Split and Dubrovnik, where the presence of green leftwing councillors is slowly transforming the political landscape and liberating space for another vision, which will create

---

*Yet, very often they are deprived of any scientific or evidence-based assessment of the climate impact on jobs in Croatia. For more information (in Croatian), see https://bit.ly/3JHFZsm.

shifts in local policies that are compatible with the objectives of just transition.

Cooperatives, on the other hand, are collaborative environments where citizens can autonomously organize and mobilize others to induce change. From this perspective, the case of ZEZ (Green Energy Cooperative*) is a true example of one of the front runners: it represents a model of citizen engagement through the expansion of renewables, mainly solar energy, in a variety of small-scale decentralized projects that cooperate with cities in Croatia and in Southeastern Europe. Their collaboration with the small city of Križevci, as well as establishing public–civic partnerships, is an adequate illustration of how collaborative practices can be employed for public interest and for supporting transformative processes.

However, while local experience and a bottom-up approach are very important courses of action, most of the workers in Croatia would be forgotten or left behind if we relied exclusively on these achievements. This is mainly because their work is still dependent on the state, and it would be misleading to deny or ignore the role of the state and the public sector – even more so since the pandemic has shown us that the state is strengthened in times of crisis.

However, as the state is often reluctant to initiate deeper systemic reform or policy shifts toward substantial democratization, we are often inclined to celebrate the small victories of the bottom-up approach. Still, this large problem is not solved by looking the other way. Public investment and the modernization of public infrastructure and services are of crucial importance for decarbonization, and they need to be coupled with maintaining high-quality jobs in the public sector. As the state has direct responsibility – as well as a mandate and the resources – to initiate this transformation now, the public sector is the appropriate terrain for political struggle that can strengthen the role of the state in providing a framework for just transition. This can also open doors to the democratization of public companies through decarbonizing work and saving jobs.

Another perspective that offers a bit more hope is the application of Kate Raworth's doughnut economics to cases of local economies

---

* Available at www.zez.coop/index_en.html.

in Croatia. Degrowth doughnut models,* developed by researchers at the Institute for Political Ecology, have been applied to a variety of Croatian cities to provide a baseline and a roadmap for how future city economies could look under ecological limitations. As well as being visually attractive, the models also provide solid orientation on how local economies can quickly transform and from where these policy shifts need to emerge.

Carriers of transition

We do not want just transition to be an artificial process. In addition to the absence of protagonists and carriers that have a stronger influence and impact, there is one more challenge for just transition that is manifested in Croatia. Ideally, we see the decarbonization of production and consumption systems as being part of an inclusive and participatory process of responding to the needs of workers and of maintaining high-quality jobs. Some initiatives are increasingly successful in decarbonizing energy production and mobility in cities, or in decreasing energy poverty, while others are successful in maintaining a high level of worker involvement in companies. Yet, in reality, this does not all happen at the same time: we need synchronicity, and for the moment it is difficult to expect it from a purely bottom-up approach. There is simply no time to waste.

Last but not least, to decarbonize our economies and keep our jobs we need to work on ensuring continuity (with the state), consensus (with movements, employers and trade unions) and speed, to achieve synchronized change. For that we need real and specific scenarios around the most carbon-intensive industries, along with feasible timescales.

Bibliography

Croatian Parliament. 2020. Adaptations to climate change in the Republic of Croatia for the period until 2040 with a view to 2070. Strategy Document 921 (https://bit.ly/3BXQNzR).

---

* See more information at http://ipe.hr/en/degrowth-donut/.

Džihić, V., and Hamilton, D. S. (eds). 2012. *Unfinished Business: The Western Balkans and the International Community*, Washington, DC: Brookings Institution Press.

Euronews. 2021. Green MP Tomasevic crushes far-right candidate in Zagreb mayoral election. *Euronews,* 31 May (https://bit.ly/3JOJqgO).

Horvat, V. 2004. Brain drain. Threat to successful transition in Balkan region? *South Eastern Europe Politics* 5(1), 76–93.

Jović, D. 2022. Post-Yugoslav states thirty years after 1991: unfinished businesses of a fivefold transition. *Journal of Balkan and Near Eastern Studies* 24(2), 193–222 (doi: 10.1080/19448953.2021.2006007).

Perspectives. 2021. Southeastern Europe: green transition and social (in)justice. *Perspectives,* Issue 9, December, Heinrich Böll Foundation (https://bit.ly/3HjBe6O).

Republic of Croatia. 2022. Conference on the Future of Europe – Croatia's transition to climate neutrality. Government Website (https://bit.ly/3giAhR9).

Just transition in the Nordic model

Simo Raittila

Just transition has already been defined at various points throughout this book, especially in the chapter entitled 'Different perspectives on just transition'. My aim in this chapter about the Nordic welfare model and just transition is to focus on the latter word in the concept: transition. I will ask the following questions. What kinds of situation are the Nordic countries transitioning from? What kinds of route are available for these countries? And what kinds of thing should be taken into account to guarantee that the transition is a just one?

In his classic work of social policy theory *The Three Worlds of Welfare Capitalism,* Gøsta Esping-Andersen (1990) defined three different types of welfare state (*regimes*). Contrary to the more modest and targeted *liberal* and family-based *conservative* regimes, the northern European *social democratic* regimes were characterized by universalistic benefits and services that aimed to provide a high standard of living and individual freedoms.

Another way of describing this is that the Nordic countries provide a higher degree of decommodification from market forces than most other countries. In a non-interventionist capitalist society labour is treated as if it is a commodity to be bought and sold like anything else. In such a society workers are unable to support themselves and their families without selling their labour to capitalist entrepreneurs. Esping-Andersen's analysis focused on people's right to a basic level of consumption, regardless of their choice of whether or not to work. A truer form of decommodification would also provide people with opportunities for self-development without limiting this to learning 'instrumental skills' that are valuable in the job market (Room 2000).

A short definition of the Nordic model, adopted in a 2021 report published by the Nordic Council of Ministers, is as follows (Alsos and Dølvik 2021):

> The models in the small, open Nordic economies are founded on three basic pillars: 1) active states with a responsible, stability-oriented macroeconomic policy, 2) strong social partners and coordinated collective bargaining, and 3) universal welfare states contributing to income security, skill formation and labour market participation. In coordination with a market and competition based business sector, the three-pillar models have helped the Nordic countries achieve a combination of efficiency and equity. The models are not static – they have been adjusted and adapted to new realities at a number of junctures. It is precisely this ability of the models to handle crises and major social changes that has been part of the success story.

As just one example of a text lauding the excellence of the Nordic model, a report published by the CMI – Martti Ahtisaari Peace Foundation explained the prosperity of Nordic countries through social policies, economic coordination and good governance that built trust between different groups and state institutions (Hiilamo and Kangas 2013).* These are also probable reasons for the countries' high placement in different rankings of well-being and happiness. The Nordic countries invest socially in individuals at different stages of their life. What is especially important, though, for the model in general and for just transition specifically, is the support provided to working-age people (Alsos and Dølvik 2021):

> For adults of prime working age, social investments also focus on further education, namely **tertiary education, life-long learning and active labour market policies**. Tertiary education deals with the creation of human capital. Life-long learning aims to up- and re-skill workers in the light of changing labour markets and technologies. The principal idea of all these policies is to employ as many people as possible. If its ability to sustain a high level of employment

---

* For another popular take on the Nordic model, see Partanen (2016).

is ignored, then the Nordic model is nothing more than a beautiful idea or abstract concept.

However, regardless of the hype around it, even the Nordic model cannot claim to represent a truly sustainable welfare state if the consumption level and well-being provided is based on excessive emissions and environmental harm. Denmark, Sweden and Finland have decreased their climate emissions from 1990 levels by around 20–30%, while the emissions of oil-rich Norway have remained pretty stable and emissions in Iceland have increased by around 30%. All of the Nordic countries apart from Sweden still have higher emissions per capita than the EU average (Council of Nordic Trade Unions, Friedrich-Ebert Stiftung and German Trade Union Confederation 2021).

The numbers above exclude the greenhouse gas emissions and sinks from the LULUCF (land use, land use change and forestry) sectors, which are especially beneficial for reducing the net emissions of Sweden, Finland and Norway, which have relatively large areas of forests that act as carbon sinks and storage. Since the forests capture carbon, this gives the countries comparatively more leeway to achieve climate neutrality – if these sinks are protected, at least. Notably, in 2021 Sweden and Finland gained notoriety in environmental circles by lobbying on behalf of their countries' forestry industries when the European Commission was working on multiple policies to protect biodiversity and to increase carbon sinks (Euroactiv 2021). According to the United Nations Framework Convention on Climate Change registry for decreasing emissions from nationally determined contributions, each country has the following current goals.

- Finland: at least 55% below 1990 levels by 2030 (updated in December 2020). (Note: the new Climate Change Act – to be decided in spring 2022 – would increase the goal to 60% by 2030, with carbon neutrality by 2035 (Finnish Ministry of Environment 2021).)
- Sweden: at least 55% below 1990 levels by 2030 (updated in December 2020). (Note: an EU-wide target via the Climate Change Act of 2017 is net zero domestic emissions by 2045.)

- Denmark: at least 55% below 1990 levels by 2030 (updated in December 2020). (Note: as above, plus the Climate Change Act aims for a 70% reduction by 2030 (London School of Economics Grantham Institute 2022).)
- Norway: at least 50%, and towards 55%, below 1990 levels by 2030 (updated February 2020). (Note: a target of 80–95% reduction in greenhouse gas emissions by 2050 according to their Climate Change Act.)
- Iceland: at least 55% below 1990 levels by 2030 (updated in February 2021). (Note: a target of carbon neutrality by 2040 according to their Climate Action Plan of 2020.)

Despite their promises to do so, not all the Nordic countries have increased their development aid to meet the common UN goal of contributing 0.7% of gross national income. Preliminary data from the Organisation for Economic Co-operation and Development (2020) shows that Finland (0.47%) and Iceland (0.29%) fall short, Denmark barely achieves the target (0.73%), while both Norway (1.11%) and Sweden (1.14%) go beyond it. Climate change will worsen the humanitarian situation worldwide and lead to an increase in people seeking refuge, which is something that the Nordic countries should take into account. Since rich countries have benefited from industrialization historically – something that has increased their emissions – they have a moral duty to act first and to finance the 'loss and damages' of climate action (and inaction) to the Global South and the most affected people and areas. Yet, the goals previously made for financing the transition were not met by the 26th Conference of the Parties in Glasgow (Tynkkynen 2021).

Like many other rich countries, the Nordics are also challenged by ageing baby-boomer populations and rising (age) dependency ratios. To simplify, there are more and more retirees and elderly people who have stopped working and who use services, and at the same time there are comparatively fewer people paying income tax, which is used to pay for these services and welfare state benefits.

The countries have had varied success in increasing immigration to account for the skew in the population pyramid. The challenge of dependency ratios has, in part, led to the increase in active labour market policies and sanctions in the countries since the 1990s. Due to such developments (and others) it has been argued that Esping-Andersen's 30-plus-year-old theory is lacking in some way.

Penttilä and Hiilamo (2017) compared the sanctioning of minimum-income schemes in twenty European countries. They followed a five-regime division (Pascual 2007) instead of Esping-Andersen's original three-regime division, and they found that while there were similarities between the Nordic countries, Finland and Norway had more lenient policies whereas Sweden had less lenient policies but gave more consideration to social workers. Denmark was not included in the study. In the context of just transition, it is noteworthy that the amount that could be sanctioned from last-resort social assistance seemed to be smaller in countries with bigger union presence and more collective bargaining.

The high degree of organization in the Nordics has meant that unions have had an important role in the development of the Nordic model and in the transmutation of productivity growth to wages and universal services. They could now play a similar role in protecting the well-being of localities and workers during the transition to a green economy.

Indeed, even though there might be some differences in how different actors perceive just transition and define its boundaries, three main dimensions can be recognized (Moodie *et al.* 2021):

- a technical dimension, which relates to the shift towards carbon-free technologies;
- a social justice dimension, which focuses on citizen involvement in the transition process, preserving jobs and protecting the most vulnerable in society from the potentially damaging socio-economic impacts of climate policies; and
- a spatial dimension, which aims to ensure that transition policies are based on territorial specificities that meet the needs of local and regional citizens.

I will argue that too narrow an understanding of all these dimensions – and especially of the social justice dimension – can limit what policies we look at. If one wants to transition with the main purposes of the welfare model intact, one has to look at social policies in their entirety.

CHALLENGING THE FOCUS ON EMPLOYMENT AND MARKET INCOME

In many chapters of this book special focus is given to how environmental action affects the workforce.* Let me cite the prior Green European Foundation framing paper on Just Transition (Franssen and Holemans 2020):

> Within the international climate community, Just Transition was increasingly framed and recognized as the trade union movement's contribution to the international climate debate. ... [In 2009] the ITUC presented Just Transition as 'a tool the trade union movement shares with the international community, aimed at smoothing the shift towards a more sustainable society and providing hope for the capacity of a "green economy" to sustain decent jobs and livelihoods for all'.
> United Nations Research Institute for Social Development (2018)

The Finnish Greens published a policy paper in early 2020 entitled 'Reilun vihreän muutoksen ohjelma', which roughly translates as 'Programme for fair green change'.† In it the party did not limit its scope to employment issues, as many much-needed policies affect consumers, and so on. This mandatory system change impacts on many aspects of society and, if not planned well enough or compensated by other policies, it would have detrimental effects on many individuals' lives and would risk increasing inequality.

The main goal of the policy paper was to provide a comprehensive overview of a potential state system that reduces emissions, environmental harm, poverty and inequality simultaneously. The

---

* See, for example, Chapter 1 of this book ('History of the just transition concept') or the prior framing paper (Franssen and Holemans 2020).

† The policy paper itself is currently only available in Finnish (see https://bit.ly/3JQfFMz). Some goals of the Finnish Greens are collected on the party's website in English at www.greens.fi/goals-and-themes/.

transition to an ecologically sustainable economy should not be allowed to lead to the poor getting poorer. People should be able to achieve a similar or better standard of living even while crises such as climate change are addressed. Some suggestions made in the policy paper included the redistribution of wealth by reforming the capital gains tax system. This would follow the model in the Mirrlees review (Mirrlees *et al.* 2011) towards freeing the normal return on capital from taxation and taxing higher capital gains more progressively, using indicators such as the genuine progress indicator and the index of sustainable economic welfare rather than relying (solely) on GDP; this also includes introducing a universal basic income and compensation, e.g. for increased transportation costs for low-income families.

Using the ideas of an active welfare state and decommodification introduced above, I would differentiate between different kinds of solution: to protect the workforce and other groups of people, and also to 'carry' individuals over the imperative transition to a sustainable system. In my view, we should ask two main questions.

1. Is the policy temporary and targeted (*interventional*) or system-wide and continual (*structural*)?
2. Is the compensation related to re-education (*capability building*) or is it monetary?

Furthermore, a judgement could be made about whether the policy is more or less active or passive. More active (labour market) policies can take more or less bureaucratic, controlling and well-received types of policy, but individual lives are clearly more entangled in such policies than in the simple exchange of money or services in kind.

Table 1. Categorizing welfare policies for just transition based on targeting and focus.

	Capability building	Monetary
Interventional	Limited time on re-education and services	Compensation of losses
Structural	Lifelong education	Tax and social policies

A stereotypical example of an *interventional policy* to tackle changing circumstances would be helping employees that are let go by a factory because environmental policies have made their product unviable. These people would be identified (targeted in an *interventional* manner) and offered services that would help them access other kinds of jobs, through re-employment or re-education (*capacity building*).

One real-life example of an interventional policy was the limited-time extra state funding (up to €8.7 million in 2020 and 2021) that was offered to the area of Jämsä in Finland. The Kaipola paper factory was closed down in 2020 and compensation packages were offered to former employees by their employer, UPM-Kymmene Oyj. However, state funding was only offered to companies, and during the Covid-19 pandemic very few larger companies invested in growth or new technologies. The mayor of Jämsä has criticized this, since there are actions that could be taken by the municipality itself, such as investing in publicly owned buildings used by multiple companies (Kotilainen 2021).

In the case of Jämsä, some of the people who lost their jobs could retain their salaries for almost half a year (January–May 2021) despite not having any more work to do. This was because of legally mandated long notice periods for employees who had worked in a company for a long time. If any unemployed person became employed or started formal education during this time, they could get the rest of the notice period's wages paid to them immediately. UPM offered the unemployed individuals education, coaching and monetary support for formal education, among other things. It promised to pay for moving costs and give the equivalent of one month's pay to individuals who moved to another UPM location, even if they started to work for a company other than UPM. The company also gave monetary support to those who started their own businesses (up to €10,000 per company).

Policies that are both *monetary* and *interventional* also benefit groups or individuals who deem a policy change to be unfair; however, instead of the welfare policies acting in line with individual needs, monetary and interventional policies only approach the problem in an impersonal manner. For example, (temporary or one-off) monetary compensation can be given to entrepreneurs whose livelihoods are destroyed by policies that, for example, ban the burning of coal or peat for energy, or that ban fur farming. One could also argue that one-time-only or limited-term policies also fall into this category. An example of such was the United States's CARES Act, which included direct cash transfers to individuals and families along with loans to small companies; this was put forward in response to the Covid-19 pandemic (Raittila and Bollain Urbieta 2021).

Social compensation can also take the form of changes to tax rates, etc., that aim to address unequal or unfair distributions of income, wealth or well-being that result from environmental policies. In this case the policies are *monetary* and *structural*. When it is not possible to identify beforehand who will be affected by a policy, or if the range of individuals is too wide, structural policies come into play more strongly, as they act more dynamically to change individual situations.

The existing social benefits, especially unemployment insurance and last-resort social assistance, are structural policies, but they are not *passive*: they are strongly tied to an individual's actions and status. In the Nordic welfare states these benefits play a large role in the current status quo. I call the policies that do not target individuals (i.e. *structural* ones) and that do not micromanage their lives 'liberating'. An example would be the idea of a permanent universal basic income. These kinds of structural and liberating policy are, in my view, most relevant to improving universal rights, social security and social services in a just transition.

Liberating policies are dynamic, and they increase the resilience of a society. For example, many states reacted to the Covid-19 pandemic with policy changes that could be interpreted as having taken (temporary) steps towards the ideals of universal basic income models.* I would mostly identify the ideal of a Nordic decommodificating welfare state with those liberating policies. I will further discuss this below when I talk about just transition and welfare state reform.

---

* For an argument that universal basic income increases resilience in such circumstances, see, for example, philosopher Otto Lehto's argument in a 2020 King's University London blog at https://bit.ly/35p3PKl.

There is a further reason not to focus solely on interventional policies that help those individuals and industries clearly identified as 'losers' from the transition. One should also ask whether policies that target these groups are fair when compared with the system-wide policies that other unemployed people face. This issue was also raised on social policy reactions to Covid-19 in Europe: people who were unemployed before the pandemic sometimes faced stricter policies than those who were unemployed just prior to it (Raittila and Bollain Urbieta 2021).

Instead of limiting the role of just transition policies (such as employment subsidies, retraining opportunities and extra compensation packages) to a targeted group of newly unemployed people or groups that are otherwise made worse off during the change (see the first row of table 1: the interventionist approach), we should also consider the more universal need for such welfare policies (see the bottom row: the structural approach). For example, support for life-long learning for all might arguably be better and fairer than ad hoc courses to individual companies' ex-employees – regardless of company size or the number of people laid off. Similarly, basic income for all who are in need is better than trying to filter out the 'unworthy'.

MARKETS AND NORDICS CLINGING TO THE PAST?

Contrary to the belief of some, the Nordic model is not detrimental to markets. Rather, Nordic countries provide highly trained workforces, social cohesion and strong regulatory structures in which corporations can thrive.* Similarly, just transition as a principle is not automatically opposed to the use of markets. Rather, it follows the same tradition of decommodification and *limited* markets that is arguably pervasive in all welfare regimes, and thus pervasive worldwide (Esping-Andersen 1990; Polanyi 1944). What we predominantly have all over the world are different degrees of mixed economies.

All Nordic countries have gone through a transition from agricultural economies to industrial ones, and then from industrial

---

* For a comparison of Denmark and the United States, see, for example, Wilkinson (2016).

to service-based economies. In the case of Finland, the share of industrial jobs decreased from 29% in 1977 to 14% in 2008 (Sutela and Lehto 2014). Since 2016, service industries have overtaken manufacturing when measured by their share of Finland's domestic value-added exports (Organisation for Economic Co-operation and Development/Statistics Finland 2020). These trends are expected to continue in the future, especially if a transition from our currently unsustainable societies is achieved.

There is a risk that in the guise of just transition, some parties will try to allocate public funds to non-green industry jobs. This was seen in the forest and paper sectors in Finland and Sweden. The forest and paper sectors, including furniture making, makes up nearly a fifth of all industrial production in Finland, according to the Ministry of Agriculture and Forestry (n.d.):

> With respect to foreign trade, forest industry is also one of the key sectors, as it accounts for a fifth of Finland's export revenues and five per cent of the gross domestic product. Relative to its size, Finland has the most forest-dependent and forest sector reliant national economy in the world.

Such dependency is not ideal if states are to refrain from investing in non-green industries. Instead, there have been calls for *divestment*. The Nordic countries have a history of state-owned enterprises (SOEs), though many have now been privatized. The Finnish state, for example, is the sole owner of Finavia, which maintains Finland's airports, and of the train monopoly VR, which was opened to provide competition for local train transport by the earlier conservative government but has recently decommissioned train carriages that could have been bought by future competitors. The Finnish state holds a major stake in the airline Finnair, the natural gas and biogas company Gasum, the oil and biofuel company Neste, and the energy company Fortum/Uniper, which, it has been claimed, is one of the EU's worst polluters (YLE 2019).

The Finnish Forest Administration (Metsähallitus) manages most of the wood supply to Finnish industries, and it contributed almost €140 million to the government in 2019. The amount of forest use is a highly contested political issue due to its effects on carbon sinks, carbon storage and biodiversity.

Here are some other examples of SOEs in the Nordics.

- In Sweden, the state fully owns Swedavia (airports) and Vatten-fall (energy), and it owns about a fifth of SAS (an airline).
- Denmark owns 29% of Københavns Lufthavne (airports) and 22% of SAS.
- Norway owns all of Avinor (airports), Gassco (which manages natural gas pipelines to Europe) and Petoro (oil and gas extraction). It also owns a majority share in the multinational company Equinor (formerly Statoil), whose main business is in petroleum and which operates in thirty-six countries around the world).

Government ownership is in itself not a problem, but due to 'ownership steering', things such as reducing employment in some of these companies, can become a politically charged topic. Also, possible reliance on income from non-green companies could mean weaker incentives for governments to make changes to their economic or competitive environment. For example, Finland was expecting a return of €2.9 billion from its SOEs – that is 4.5% of the total income in the government's original budget for 2021.

A meaningful counterexample from Sweden is the company Sam-hall, which creates jobs for people with disabilities. Despite criticism of some of its functions, a similar limited 'employer of last resort' is being introduced in Finland. Also, Norway owns many companies aiming to use the money gained from the country's massive fossil fuel industry to find more sustainable alternatives.

There should be sufficient space for *creative destruction*: that is, for old technologies, industries and companies to be replaced with new, ideally greener, ones.* Certainly in Finland, many of the interventions and investments made by the government fail to

---

*For an in-depth discussion on creative destruction and government policies see, for example, Aghion *et al.* (2021). For a discussion on creative destruction and sustainability in the UK and Finland, see Kivimaa and Kern (2016).

follow this ideal and instead subsidize current industries in place of future ones.

WELFARE STATE REFORM

It should be recognized that much of the value generated in modern economies comes not from material production but from the service and information sectors of the economy. New industries and new forms of work challenge existing welfare systems, which calls for a more 'back to basics' approach according to the report 'The future of work in the Nordic countries' (Alsos and Dølvik 2021):

> Rather than presenting radically new proposals, we promote a 'back to basics' approach where the foundational pillars of the Nordic models are strengthened to meet the future of working life. In some areas this may be done via a more visible government hand and less leeway for the 'invisible hand' of market forces in labour policy; however, at an overall level we believe the key to mastering the transition to the future of working life lies in further developing and vitalising the partnership between the social partners and the state centrally as well as locally. In parts of the private service sector, this will likely require public actions to stimulate increased organisation.

Following on from the above, I would argue there is a need to broaden the discussion on just transition. Rather than mostly focusing on local and occupational interventions, a wide range of universal policies should be put in place.

Social scientists Merrild and Andersen (2019) have argued that the neoliberal shift in Danish welfare policies has led to 'new forms of uncertainty' for those depending on the services of the welfare state, as well as an expectation of 'sameness'. This masks individual suffering and challenges that do not fit the cultural moulds of either those individuals fit to work (and mandated to do so) or those clearly deserving support. This is a risk that is also heightened with more targeted ('upper row') just transition policies if the old Esping-Andersenian goal of decommodification is given up in the

Nordic countries. One could claim that it has already been, slowly and partially, chipped away.

The Finnish research group BIOS (2019) has published an extensive report on *ecological reconstruction*. In it they argued that the 'war metaphor' of reorganizing an economy in response to the climate and biodiversity crises should be replaced with a 'post-war reconstruction metaphor': an increase in democratic power. This participatory practice is also at the core of just transition policies (a *social justice dimension* and a *spatial dimension*). As people's jobs and livelihoods are on the line, Nordic governments need to be able to provide more robust guarantees of basic needs being met and of rampant inequality being addressed. One debate is whether a job guarantee policy, recommended by BIOS among others, is a good way to go, or if a universal basic income is a more just and dynamic policy:

> Within job guarantee, the public sector offers jobs to all willing employees with salaries that, in practice, become the minimum wage. The jobs do not require long training but have decent conditions and are directed toward improving the society. The government finances the guarantee, but the jobs may be organised more locally – for instance, at the municipal level.

Instead of just looking at where we currently are and how we should achieve a just transition, we should aim to define our new ideals, i.e. where we want to end up. In my view, this should include a discussion about the degree of decommodification and a renewal of the ideals of the Nordic welfare model.

Bibliography

Aghion, P., Antonin, C., and Bunel, S. 2021. *The Power of Creative Destruction: Economic Upheaval and the Wealth of Nations*. Cambridge MA: Belknap/Harvard University Press.

Alsos, K., and Dølvik, J. E. 2021. The future of work in the Nordic countries: opportunities and challenges for the Nordic working life model. Report, 6 May, Nordic Co-operation (https://bit.ly/3MgZFW7).

BIOS. 2019. Ecological reconstruction in Finland. Website, 8 November, BIOS (https://eco.bios.fi/). (In Finnish, but also available in English.)

Council of Nordic Trade Unions, Friedrich-Ebert Stiftung and German Trade Union Confederation. 2021. The road towards a carbon-free society – a Nordic–German trade union cooperation on just transition. Report, accessed via NTU website (https://bit.ly/3pyxnNc).

Esping-Andersen, G. 1990. *The Three Worlds of Welfare Capitalism*. Cambridge: Polity Press.

Finnish Government. 2020. Ministeri Tuppurainen: valtionyhtiöiden osinkoja tarvitaan hyvinvointivaltion rahoittamiseen. Press Release, 8 October (https://bit.ly/34IsIk9) (In Finnish).

Finnish Ministry of Environment. 2021. New climate change act sent out for comments. Press release, 2 July, government website (https://bit.ly/3KalG79).

Franssen, M. M., and Holemans, D. 2020. Climate, jobs and justice for a green and socially just transition. Working Paper, December (https://bit.ly/3vQJxoD).

Hiilamo, H., and Kangas, O. 2013. Recipe for better life: experiences from Nordic countries. Report, October, Center for Crisis Management (https://bit.ly/3K44WOS).

Just Transition Research Collaborative. 2018. Mapping just transition(s) to a low-carbon world. Research Report, UNRISD (https://bit.ly/3Jbm3ON).

Kivimaa, P., and Kern, F. 2016. Creative destruction or mere niche support? Innovation policy mixes for sustainability transitions. *Research Policy* 45(1), 205–217.

Kotilainen, V. 2021. Hallitus lupasi Kaipolan tehtaan menettäneelle Jämsän seudulle miljoonien elpymistuet, mutta korona vei yritysten kyvyn investoida. *yle*, 5 March (https://bit.ly/3pC4wY7) (In Finnish).

Lehto, O. 2020. Universal basic income as a crisis response to the novel coronavirus. Blog Post, 1 May, Kings College London (https://bit.ly/36Oh1db).

London School of Economics Grantham Institute. 2022. Climate laws of the world: Denmark. Website (https://bit.ly/3MjLhfx).

Merrild, C., and Andersen, R. S. 2019. Welfare transformations and expectations of sameness. Living on the margins in Denmark: perspectives on social differences in the welfare state. *Nordic Journal of Social Research* 10(1), 66–85 (doi: 10.7577/njsr.2858).

Ministry of Agriculture and Forestry of Finland. n.d. Forest industry in Finland. Website (https://bit.ly/3vEVHR8).

Mirrlees, J., Adam, S., Besley, T., Blundell, R., Bond, S., Chote, R., Gammie, M., Johnson, P., Myles, G., and Poterba, J. M. 2011. *Tax by Design*. Institute for Fiscal Studies and Oxford University Press.

Moodie, J., Tapia, C. Löfving, L., Sánchez Gassen, N., and Cedergren, E. 2021. Towards a territorially just climate transition – assessing the Swedish EU territorial just transition plan development process. *Sustainability* 13(13), Paper 7505.

Organisation for Economic Co-operation and Development. 2020. Official development assistance (ODA). Website ⟨https://bit.ly/36OdZFa⟩.

Organisation for Economic Co-operation and Development and Statistics Finland. 2020. Globalisation in Finland: granular insights into the impact on businesses and employment. Report ⟨https://bit.ly/3tncIg5⟩.

Partanen, A. 2016. *The Nordic Theory of Everything: In Search of a Better Life*. New York: Harper Collins.

Pascual, A. S. 2007. Activation regimes in Europe: a clustering exercise. In *Reshaping Welfare States and Activation Regimes in Europe*, edited by Pascual, A., and Magnusson, L. Brussels: P.I.E. Peter Lang S.A.

Penttilä, R., and Hiilamo, H. 2017. Toimeentulotuen saajien sanktiointi eurooppalaisessa vertailussa. *Yhteiskuntapolitiikka* 82(4) ⟨https://bit.ly/3pynmiM⟩ (In Finnish).

Polanyi, K. *The Great Transformation: The Political and Economic Origins of Our Time*. Boston, MA: Beacon Press.

Raittila, S., and Bollain Urbieta, J. 2021. Steps towards universal basic income? The effect of the Covid-19 crisis on welfare policies and support for UBI in the European Union. Research Report, Green European Foundation ⟨https://bit.ly/3HG6tK7⟩.

Room, G. 2000. Commodification and decommodification: a developmental critique. *Policy and Politics* 28(3), 331–351. (doi: 10.1332/0305 573002501009).

Sutela, H., and Lehto, A.-M. 2014. Työolojen muutokset 1977–2013. Report, Statistics Finland ⟨https://bit.ly/3tKbY4X⟩ (In Finnish).

Tynkkynen, O. 2021. Progress at Glasgow Climate Conference on three fronts. Blog Post, Finnish Innovation Fund Sitra ⟨https://bit.ly/3ttPxAA⟩.

UPM-Kymmene Oyj. 2020. UPM tukee Kaipolan henkilöstöä laajalla muutosturvapaketilla. Website, 15 October ⟨https://bit.ly/35slWzF⟩ (In Finnish).

Vanttinen, P. 2021. Environmentalists 'up in arms' about Finnish–Swedish defence of forest industry. *Euractiv*, 31 May (https://bit.ly/36MIlb5).

Wilkinson, W. 2016. The freedom lover's case for the welfare state. *Vox*, 1 September (https://bit.ly/3hcSa4k)

YLE. 2019. Fortum/Uniper among Europe's worst polluters, say NGOs. *yle*, 8 April (https://bit.ly/3sBYRTM).

The uneven rise of just transition in Western Europe

Dirk Holemans and Elina Volodchenko

This chapter focuses on Western Europe. In this part of Europe, countries such as Belgium, Germany and the Netherlands are currently working towards an economic transition and looking at how to include the principle of just transition, to a greater or lesser extent. We will concentrate on industrial policies in Belgium (specifically Flanders, which has the second largest petrochemical cluster in the world), Germany and the Netherlands, and we will consider how societal debates on this necessary transition have been organized. We will focus on what these countries are currently doing to support a just transition. The economies comprise many sectors, of course, from coal mining to the chemical industry and car manufacturing. In this sense, we do not aim to provide an exhaustive overview; rather, we want to point out relevant developments.

HEAVY INDUSTRY IN WESTERN EUROPE

The decarbonization of heavy industry is a crucial step towards realizing a carbon-neutral Europe by 2050. Currently, the production of iron, steel, cement and aluminum leads to a substantial proportion of emissions (Bollen *et al.* 2021). These emissions are concentrated in a few firms and regions. Thirteen major industrial firms in Belgium and thirteen in the Netherlands account for more than 70% of all industrial emissions from the EU emissions trading system. In the Netherlands, thirteen entities (only 2% of the total number) account for 79% of total emissions. In Germany, fifty-five major firms (6% of the total) account for 65% of total emissions. The firms involved include ArcelorMittal, Shell, BASF, Air Liquide and Heidelberg Cement.

Furthermore, some regions will be affected more than others, because industries and hard coal regions are not evenly distributed. Mines are concentrated in certain places, and most high-emitting industries in Western Europe are clustered. The conversion of the coal mine industries represents one of the biggest challenges for these regions. In some areas the benefits will not be felt immediately, which can lead to substantial protests if the issue is not dealt with in the right way. Job losses and a reduced quality of life are the main fears that can lead to resistance. Another fear is the perceived unfair distributional costs of the transition. The task is now to listen to these concerns and to provide a voice for all the stakeholders involved, in order to create a just transition that leaves no one behind. Unions and civil society organizations stress that strength is to be found in the inclusion of the civil society and the affected communities as well. This needs to be supported by bottom-up initiatives. The case studies presented will show how important the participation and inclusion of the civil society is in shaping a just transition pathway for the coming years.

Germany

Germany has a couple of clustered coal regions. For many years the Ruhr and Saarland areas have been the main hard coal regions (Dahlbeck and Gärtner 2019). Germany also has four main lignite mining districts: the Rhineland, Helmstedt, Central Germany and Lusastia. The country has a long history of lignite and hard coal mining. Hard coal was mainly mined in West Germany, while lignite played an important role in both West and East Germany. The hard coal mines in Germany were very labour intensive due to the underground work that had to be done. At the end of the 1950s, for instance, nearly 600,000 people were employed in the coal mines.

As a result of industrialization, the Ruhr area became one of the country's most densely populated regions. Almost every tenth inhabitant of the region was employed in the mining sector. The area went through structural policy developments for decades. The process of phasing out coal started with a crisis in the 1960s (Anczewska *et al.* 2020). Firstly, coal from the Ruhr area was no longer competitive on the international market, as oil was gaining international importance. Secondly, the hard coal in the Ruhr area lay at much deeper levels than in other regions, which presented more difficulties: the production costs were higher and it was more time-consuming to mine. These economic matters caused the move towards a transition. By 1963, thirty mines had already been closed, and these were followed by numerous other closures in the following decades. The first coal crisis in the 1960s cut the number of available jobs in half: the number of employees fell from 495,800 in 1957 to 210,300 in 1968 (Dahlbeck and Gärtner 2019). Coal mining declined further over the years. After the 2000s the number of employees decreased even more, from just 45,400 in 2000 to 5,800 in 2016.

For a long time, politicians tried to maintain the competitiveness of the Ruhr area's coal (Anczewska *et al.* 2020). A couple of measures were taken in an attempt to realize this. The first was the founding of the Ruhrkohle AG, an initiative that merged the mines together. Another measure was the introduction of the 'coal penny', where energy supply companies were obliged to prioritize German coal over imported coal. This was stopped by the Federal Constitutional Court in 1995. Instead, direct subsidies were given, but politicians were trying to save something that simply could not be saved anymore, when instead the money could have been invested earlier into new sources of energy. The delay in making the necessary changes (of an inevitable transition) raised costs considerably. (What is more, the Ruhr area was still receiving coal subsidies until 2007.)

The culture, the region and the people were closely intertwined in the romanticization of coal mining (Anczewska *et al.* 2020). Mining was more than a job for many: it carried the spirit of the whole community. Many therefore feared the transition in the region – something that sparked massive protests and created social disruption. To cushion the phase out of coal, subsidies were provided. This made the early retirement of many miners possible. In the beginning, many miners were also given a new job in the steel and automobile industries, but this proved to be unsuccessful.

Eventually, the hard coal and steel sector in Germany accelerated into a bigger structural crisis. In 1968 a structural policy was introduced: the Ruhr Development Programme was started (Dahlbeck and Gärtner 2019). This programme was then transferred to the North Rhine-Westphalia Programme in 1970. It was the first time an integrated approach of different measures was used. The focus of these measures was both the socially responsible reduction of employment in the coal industry and the expansion of infrastructure in the region. So, by expanding the public transport system, the road network and research and education infrastructure could be expanded as well. For example, until the 1960s no university had existed in the Ruhr area.

In the 1970s a transition was being made to a service society in Germany (Dahlbeck and Gärtner 2019). The share of employment in the service sector has risen since then, while the share of employment in the manufacturing sector has declined. These trends can also be seen both in the federal state of North Rhine-Westphalia and in the Ruhr area. In the latter, the transition to a service-based economy only began to develop three years later, because of the greater importance of the coal and steel industry in this region. However, due to lower demand for labour in the service sector, not all job losses could be compensated for. Moreover, industrial workers often only partly fit the requirements of the new service jobs – that is why re-education and investment in education were such important measures.

In the southern Ruhr area, the decrease in coal and steel jobs was further compensated for by expanding universities in Bochum, Dortmund, Duisburg-Essen and other established technology centres (Dahlbeck and Gärtner 2019). Alongside these institutions, high-quality services and cultural centres have emerged in these regions, while some former industries were converted into accommodation and cultural centres (Anczewska *et al.* 2020). The region underwent a transition that hardly anyone had thought was possible. The state invested more in what has now become a knowledge-based and touristic economy (Just Transition Research Collaborative 2018).

The last hard coal mines had closed by the end of 2018 (Just Transition Research Collaborative 2018): a significant example of how a just transition can be realized. Although a lot of people initially lost their jobs, the transition was stretched over a long period of time – the phase out of coal in the Ruhr area took sixty years. (Of course, we do not now have that much time left for the transition of other regions.) Trade unions were strongly involved in shaping the social implications, ensuring that there were social safety nets. However, since unemployment rates are still high, it is clear that the transformation in the Ruhr area was not perfect. The hard coal transition was an example of a substantial but not completely successful approach to a just transition. If looked at in a narrow way, it provided working solutions for the affected workers. However, if one looks at things more broadly, assessing other factors such as identity and self-esteem, the results were less positive.

Recently, Germany's biggest focus is on the coal and lignite in Lusatia. In 2018 a Coal Phase-Out Commission was formed that brought together different stakeholders, such as trade unions, industries, coal regions, non-governmental organizations (NGOs), research institutes and representatives of the affected communities (Reitzenstein and Popp 2019). The Commission itself consists of twenty-eight members. Involving a civil society is often overlooked, but here the members convey the perspectives of people and communities in the affected coal regions (Anczewska *et al.* 2020). This participation did not play a significant role in the transition of the Ruhr area, for instance, since involving the civil society was institutionalized only fairly recently. For example, in Lusatia the Citizens' Region Lusatia was initiated in order to actively represent the perspectives of the people and the communities within the region.

The German experience with the Coal Phase-Out Commission shows that multi-stakeholder commissions can play a big role in facilitating a just, sectoral transition (Reitzenstein and Popp 2019). They can lay out an important foundation for eventual policy pathways, but they cannot replace political leadership, which also needs to happen. Trade unions have been actively involved as well: a metalworkers' union, IG Metall, came forward to state the need for more ambitious climate interventions, and it is now joining forces with the climate movement (Bollen *et al.* 2021). The Commission has formulated a goal of phasing out coal by 2038 at the latest. It has also determined accompanying economic, social and structural support measures.

In 2020 the German parliament voted for the Coal Phase-Out Law, which regulates a phase out pathway and stipulates 2038 as the end date (Heilmann and Popp 2020). It also provides compensation for the hard coal phase out, and it provides transition payments for the employees of the regulated closures. The law also includes a ban on new coal power facilities. In addition, the structural change law will regulate transition support (mainly financial) for the three lignite regions. In total, a budget of €40 billion is available. Although this represents a substantial achievement, the law has been criticized by climate movements because 2038 is not in line with the Paris Agreement, whereby all countries in the Organisation for Economic Co-operation and Development need to phase out coal by 2030.

Belgium

Belgium, and more specifically its coastal region Flanders, has one of the biggest petrochemical industries in Europe. It is mostly divided into clusters. For instance, the port of Antwerp is home to the largest integrated energy and chemical cluster in Europe. Steel, refineries and the chemical industry are the basis of the Flemish economy. In 2019, 6.3 million tonnes of steel was produced in Flanders alone (Deloitte 2020). The Paris Agreement is forcing industrial sectors like these to undergo drastic transformations towards carbon neutrality.

With the enormous presence of intensive industry in Flanders, one would expect the Flemish government to take bold action and build a broad support base all over society. But the reality is sobering: while two relevant initiatives have been undertaken, compared with neighbouring countries such as Germany and the Netherlands, Flanders is lagging behind (Beys 2020). In recent years Germany and the Netherlands have taken up climate plans for 2050 that have even been made into law. Clear targets are being set, and these are coupled with action plans and sources of funding. However, climate ambition is missing in Flanders, and just transition is not yet on the table. It is, for instance, no coincidence that in June 2021, slightly before launching the European Commission's Fit for 55 climate package, Frans Timmermans, in an interview on the Flemish public news channel VRT, called on Flanders to be more ambitious. But the Flemish government thereafter refused to support the Fit for 55 programme, because climate action should be 'feasible and affordable', as the Flemish minister of climate stated.

So what are the two steps that Flanders has already taken towards a new industrial policy framework? First, in 2019 the (previous) Flemish government launched its 'Moonshot' programme, which provides an annual subsidy of €20 million for twenty years (Bollen *et al.* 2021). The Moonshot programme focuses on four different research trajectories (Deloitte 2020). The first deals with bio-based chemistry, as an alternative way to use renewable and CO_2-friendly raw materials. The second focuses on circularity and carbon in materials. The third is focused on researching electrification and on a radical transformation of the associated processes to make them more carbon smart. The last trajectory focuses on energy innovation.

While the Moonshot programme is welcome, two points of criticism are relevant. Firstly, the budget is very small compared with the challenges involved. Secondly, the programme does not focus on the development of demonstration and commercialization pilots. In other words, there is no coherent set of support policies that bridge the different steps towards industrial-scale production large enough to supplant current (fossil/linear) production routes.

As a second step, in 2020 a roadmap of potential technological trajectories for an industrial transition towards carbon-circular and low-CO_2 Flemish industry was published (Deloitte 2020). To explore the possibilities, the government asked external partners to undertake an in-depth analysis to study the possible pathways. The resulting roadmap was made on the basis of a study conducted by external consultants in direct cooperation with the industries. However, neither trade unions nor environmental organizations were involved in any meaningful way. The plans for 2020–2030 are mainly focused on further research and development and on the realization of pilot projects for new technologies, next to electrification, a circular economy and green hydrogen. As is the case for most similar scenarios, the assumption is that new technologies will be available from 2035. We expect the period between 2035 and 2050 to involve innovative CO_2-reducing technologies, but this is just an assumption, and the uncertainties are still big.

An important additional part of the roadmap is staying competitive while becoming CO_2 neutral. The roadmap is very much focused on potential technological trajectories, while remaining growth-focused and competitive with neighbouring countries. It does not explore any other scenarios; the scope of the study is largely focused on the technological aspects and pathways for an industrial transition. More precisely, it does not consider demand reduction in certain sectors. For example, it takes 'business as usual' as the baseline in chemical and plastic production – it even forecasts growing export volumes. It is clear that the measures are still in line with the old paradigm: competitiveness is still the focal point, and climate policies are primarily shaped to fit the economic standards. This could stand in the way of a real climate transformation.

In conclusion, one can state that the Flemish government does indeed have a roadmap for the future of its industry, but it is one that is heavily technology-driven and that neglects civil society and the challenge of a just transition. The latter concept was not even mentioned when presenting the roadmap. How these first (and cautious) steps are going to be converted into more ambitious industrial policies in the upcoming years remains an open question. Thus, the reality is rather sobering. Firstly, there is not even a mention of dedicating a specific ministry or fund to just transition, as is the case in Spain, for example. Secondly, the Flemish government and the authority of its biggest port, Antwerp, are betting on maintaining their high-emitting industries while capturing the CO_2. So, in a 2021 communication from the port authorities, the headline looked quite promising: halving CO_2 emissions by 2030.* But the goal is not to reduce the actual *production* of CO_2, but to capture and store it, while looking for possibilities for reuse 'against reasonable costs'. The main goal is to explore 'the technical and economic feasibility of CO_2 infrastructure to support potentially future CCUS (Carbon Capture Utilisation & Storage) applications'. Thus, the focus is on the capture and storage of CO_2, and, in the long term, the reuse of CO_2 as a raw material. Through these technologies, which are still to develop, the hope is for a 'transition to a carbon-neutral port'.

---

* Available at https://bit.ly/3BLgeVg.

One important note here is that Belgium does not have the subsurface suitable for carbon storage. We are therefore talking about the need to transport CO_2 across borders and to store it permanently in empty foreign gas fields under the sea.

Trying to maintain the existing energy and chemistry cluster – with its well-paid jobs, of course – seems to mean that dicussions about a just transition of these sectors are not required. But it is a risky strategy, because when it fails and when industries in other countries develop a biobased industry, for instance, no longer relying on the use of fossil fuels, one could talk about the risk of a 'Kodak moment': that is, a disruptive change that causes very hard times for supposedly strong companies.

The fact that the Flemish government has not really involved civil society in a structural way in developing its industrial policy is a remarkable and regrettable change from the political past. This is because Belgium still has a strong civil society. And moreover, the trade unions and the environmental movement there have also been working together for a long time. As far back as 1985, the working group Arbeid & Milieu ('Labour & Environment') was established (Reset.Vlaanderen 2021). Trade unions and environmental movements felt the need to meet up more regularly to discuss topics that interlinked labour and the environment. The working group was soon turned into an NGO as topics such as climate change gained more and more relevance. The initial objectives focused on consultation, research, spreading information and raising awareness. Gradually, new organizations such as Civil Society Network Transition (Transitienetwerk Middenveld) also started taking on these tasks. In the meantime Arbeid & Milieu kept focusing on the transition to a just and green economy.

Eventually, Arbeid & Milieu merged with Transitienetwerk Middenveld to form what is now called Reset.Vlaanderen. Its mission is to work together with different partners and stakeholders in Flanders (such as the unions and the environmental movement, but also other big civil associations) to accelerate the transition towards an ecologically and socially just Flanders. The aim is to combine social justice with climate policy. It is a unique organization in Europe, where people actively seek out discussion with multiple stakeholders.

Close collaboration between trade unions and environmental movements does not happen very often. Reset.Vlaanderen really carries the fundaments of a just transition in its origin and development, as coordinator Vanya Verschoore laid out in a May 2021 interview:

> The consultation platform originated in the early years of Arbeid & Milieu. The environmental movement had not been around that long and at that time (in the 1970s and 1980s) there was no consultation platform to discuss climate issues with the trade unions. The actors did not know each other, so it was necessary that a platform was created. It is also needed today, but the role of Reset.Vlaanderen is different. It is no longer about getting to know each other and exchanging ideas. The transition really needs to be accelerated and action needs to be taken more quickly. That is a big difference with 30 years ago. At that time, the situation was not considered so precarious. At the moment, some (industrial) sectors are on the verge of disappearing. Just talking about this will no longer be enough, we really need to work together to put plans on the table for those sectors. But as an organization we are unique in that, Belgium is unique in that respect. Reset.Vlaanderen is really structured in order to provide a consultation platform.

The Netherlands

The Netherlands offers an interesting example of a government building on a tradition of consultation with all social actors to organize a broad public discussion about how climate objectives could be reached in an economically and socially sound way. In its Coalition Agreement of 2017, the then Dutch government announced the establishment of a Climate Agreement, organizing so-called Climate Tables in 2018 to this end. No fewer than 150 organizations sat together and worked on proposals that would ensure that the Netherlands emits half as many greenhouse gases in 2030 as it did in 1990. At five Climate Table sessions they discussed electricity, industry, the built environment, agriculture, and land use and mobility.*

---

* Available at https://bit.ly/3JfVFCH.

The Climate Tables represented an important initiative, but they did not go smoothly. At the end of the process, both the environmental organizations and the largest union walked away from the discussions. Their main objection was that there would be no general CO_2 tax for industry. Thus, while sparing large companies, private households would be forced to pay too much. On the basis of the Climate Tables, a Climate Agreement was nevertheless made: a package of measures and agreements between companies, civil society organizations and government to jointly halve the emission of greenhouse gases in the Netherlands by 2030.[*] Meanwhile, it appears that neither the Climate Tables nor the Climate Agreement offered a guarantee of a decisive climate policy, let alone one that included the principles of a just transition. The Netherlands Environmental Planning Agency concluded in 2020 that the climate targets will not be achieved by 2030 and that the government must accelerate.

However, this does not imply a complete standstill in the Netherlands. On the contrary, specific cooperation between the environmental movement and the trade unions, focusing on the future of the steel giant Tata Steel, is unique.[†] These social actors joined forces with the goal of making the factory carbon-neutral (Fonteyn 2020). Tata Steel is one of the largest greenhouse gas emitters in the Netherlands, so this agreement was a large sustainability challenge. What is remarkable is that the largest trade union in the Netherlands, the Dutch Trade Union Federation (FNV), is working with the steel factory to plan the transition towards climate neutrality by 2050.

The working group Zeester planned an alternative route to achieving green steel. The green movement proposed a more cost-effective climate-friendly alternative to carbon capture and storage. With this input, they decided the best option was to follow a green hydrogen track to run steel production instead. They planned this alternative route to save 5 million tonnes of CO_2 by 2028, while also benefiting from a direct reduction in environmental pollution and health damage. Using hydrogen and renewable energy to produce steel instead of using carbon capture is an investment that will be

---

[*] Available at www.klimaatakkoord.nl.
[†] Available at https://bit.ly/3Ii56S7.

better in the long run. The jobs are not lost but will be more sustainable in the future. The green movement wrote a letter to the Ministry of Economic Affairs and Climate Policy in the Netherlands, proposing the Zeester working plan with the alternative route to achieving green steel. They concluded that steel production could be run on green hydrogen as early as 2030. They demanded the state to take a proactive role. In the meantime, the environmental organizations asked the trade unions to enter these discussions as well. Eventually, all parties seemed to be on the same wavelength about the idea: the trade unions, the environmental movement and the government, as well as Tata Steel. The Zeester plan was brought into fruition.

However, here again, reality is unrulier than expected. In September 2021 there was finally a parliamentary debate on how to 'green' the steel factory: either by capturing the CO_2 that is emitted or by using green hydrogen. Progressive political parties and trade unions were pushing for the latter route, but the answer of the minister responsible, Stef Blok, was beyond sobering. His main message was that ultimately Tata Steel itself is responsible for greening its production processes: heavily polluting companies had to make their own considerations and choices, and this was not the responsibility of the government. This is a sad example of how the politician in charge neglected the fruits of precious social cooperation.

OVERVIEW

One would expect that Western European countries – with their welfare states (despite erosion from austerity policies), their concentration of high-level knowledge and their industrial centres and financial resources – would already have taken the lead in implementing transformative economic changes, also incorporating the principle of just transition. This is sadly not the reality. Although some relevant experiments are underway (see the previous discussion about how the transition of certain regions has proven possible and Germany has decided to phase out coal mining), one can still observe the potential risk of what is called a 'lock-in' in the economic literature. The current dominant industries, although under pressure from global developments such as climate agreements and also local, niche alternative developments, are trying to survive by making their processes and products better instead of making the radical choice of developing totally new, better processes and products. So, in theory, everybody is in favour of a circular economy, and yet we remain for the most part stuck in the traditional, linear take–make–use–dispose economy. This is clear in the chemical industry, which is still investing to produce more plastic disposables.

Furthermore, the necessary increase in climate policy ambitions – think of the Fit for 55 package – will only make the challenge bigger: transforming industries over an even shorter period due to the fact that substantial time has been lost by being attached to the old paradigm. A clear example is the car manufacturing industry, which is still very strong in Germany. As the diesel scandal showed, the automobile industry has until recently put most of its energy (at the policy level) into lobbying for lighter regulation on car emission standards, even unafraid of implementing fraudulent practices. And although since the scandal they have introduced investment programmes towards electrification, valuable time has been lost in defending and using twentieth-century technology (e.g. combustion motors running on fossil fuels) rather than taking the lead in developing and producing electric cars. Not only this, but also, now that this transformative change is evitable, many other problems connected with just transition come to the fore. For example, the production of electric cars needs fewer components and less labour, and also the production of batteries is largely automated. This puts trade unions in a difficult position: defending the rights of workers while at the same time acknowledging that we are running out of time in view of climate disruption (Galgóczi 2019).

In the light of this sober analysis, we have to acknowledge the once-in-a-decade change that we must harness for a positive transition. The Covid-19 crisis has shown that democratic governments are able to steer the economy if they have the guts to do so, and the EU is mobilizing immense funds to rebuild society after Covid-19. It would be unthinkable that the countries of Western Europe, with their financial, economic and social capital, would use their gigantic budgets only for greening the old industries instead of investing in the bold changes we owe to our children and to future generations.

This is not just a radical statement – it is a necessary conclusion, without naivety. One has to acknowledge that the insufficient progress that has been made in the heavy industries is partly due to the fear of competitive pressures internationally. The discourse in the different case studies, including government policies and proposals, remains heavily focused on competitiveness. The national governments and industries still want to, and feel obliged to, stay competitive on a global scale and to focus on optimizing economic growth and profit. If possible, they want to postpone transformative changes by betting on the capture and storage of greenhouse gases. Their reasoning is that they operate in a global context and therefeore have to compete with other companies that are currently not bound by CO_2 emissions trading laws or strong labour laws. This is an understandable concern, of course: the free flow of capital has already led to increasing production in other parts of the world. If the EU alone introduces strong climate policies, the danger of so-called carbon leakage is real: EU-based companies could move carbon-intensive production abroad to take advantage of lax standards, or EU products could be replaced by more carbon-intensive imports. The fundamental solution is the creation of a level playing field. Therefore, the proposed Carbon Border Adjustment Mechanism (CBAM), or cross-border CO_2 tax, is an urgent policy measure. The CBAM can effectively prevent the risk of carbon leakage and dismiss the narrative that the EU's increased ambition on climate mitigation would harm European industries.

TO CONCLUDE – THE FUTURE IS IN THEIR HANDS

While putting the finishing touches to this chapter, a new German government, with clear ambitions, was formed. The coalition agreement builds on the objective of limiting global warming to 1.5 °C. This focus on climate by the new government is reflected, among other things, in the merging of the competences around the climate and the economy in one 'super ministry' under the leadership of the Green Party minister Robert Habeck. This new traffic light coalition supports the Fit for 55 package and has formulated many concrete objectives and measures. For example, substantial investments are being made in renewable energy: by 2030, it must provide 80% of electricity demand. Also by 2030, half of the energy for heating must be renewable as well. Moreover, the government wants citizens and municipalities to participate even more strongly in local renewable energy.

Another milestone is 'coal exit': the end date is now 2030. By bringing forward the closing date by 8 years, Germany will avoid producing about as much CO_2 as the annual emissions of the Netherlands from 2030 onwards. In 2022, a step-by-step plan will follow to make this goal a reality, including social support and investment in the affected regions.

It is striking how these ambitious climate policies are combined with social policy, including, for example, investment in public transport and social housing. The government will launch proposals to distribute the CO_2 cost more equitably between tenants and landlords, and it will come up with a social mechanism to protect vulnerable groups from rising CO_2 costs. Add to this the reaffirmation of the just transition policy and support for regions with a coal industry, and we can certainly look forward to the roll-out of this ambitious coalition agreement. It will hopefully inspire its neighbouring countries to set the same ambitions and a just course of action.

Bibliography

Anczewska, M., de Grandpré, J., Mantzaris, N., Stefanov, G., and Treadwell, K. 2020. Just transition to climate neutrality. Doing right by the regions. Report, February, World Wildlife Fund, Germany (https://bit.ly/3Md5MdU).

Beys, O. 2020. Nieuwe studie legt eerste bouwsteen voor industrieel klimaatpact. Bond Beter Leefmilieu, 12 November (https://bit.ly/3pzapW3) (In Dutch).

Bollen, Y. Hauwaert, T., and Beys, O. 2021. For a fair and effective industrial climate transition: support measures for heavy industry in Belgium, the Netherlands and Germany. Working Paper, August, European Trade Union Institute, Brussels (https://bit.ly/37YzQdL).

Dahlbeck, E., and Gärtner, S. 2019. Just transition for regions and generations. Experiences from structural change in the Ruhr area. Report, January, World Wildlife Fund, Germany (https://bit.ly/35swM8P).

Deloitte. 2020. Naar een koolstofcirculaire en CO2-arme Vlaamse industrie. Report, November, Studie in opdracht van het Agentschap Innoveren & Ondernemen (https://bit.ly/3tulFUC) (In Dutch).

Fonteyn, J. 2021. Milieubeweging en vakbond slaan handen in elkaar: Tata Steel op weg naar klimaatneutraliteit. Bond Beter Leefmilieu, 25 June (https://bit.ly/3rzoowi) (In Dutch).

Galgóczi, B. (ed). 2019. *Towards a Just Transition: Coal, Cars and the World of Work*. Brussels: European Trade Union Institute.

Heilmann, F., and Popp, R. 2020. How (not) to phase-out coal: lessons from Germany for just and timely coal exits. Briefing Paper, September, European Climate Initiative (https://bit.ly/3MlKTxa).

Just Transition Research Collaborative. 2018. Mapping just transition(s) to a low-carbon world. Report, 28 November, JTRC.

Novy, A. 2020. The political trilemma of contemporary social-ecological transformation – lessons from Karl Polanyi's *The Great Transformation*. *Globalizations* 19(1), 59–80 (doi: 10.1080/14747731.2020.1850073).

Reitzenstein, A., and Popp, R. 2019. The German Coal Commission – a role model for transformative change? Briefing Paper, April, E3G (https://bit.ly/3Mmu3OX).

Reset.Vlaanderen. 2021. Reset.Vlaanderen. Blog Post (https://reset.vlaanderen/) (In Dutch).

Wehrmann, B. 2020. Bumpy conclusion of Germany's landmark coal act clears way to next energy transition chapters. *Clean Energy Wire*, 3 July (https://bit.ly/3pyerhx).

PART III

Just Transition and the EU Green Deal

The European Green Deal as the new social contract

Sara Matthieu

THE GREEN DEAL: A RADICAL DEPARTURE

In assessing how the idea of a just transition plays out at the European level, it is helpful to recall and put into context the major policy shift of the EU in recent years. In 2019, the European Parliament gave the Commission a mandate to pursue the Green Deal in order to transform Europe into the first climate-neutral continent. This is a radical departure from the policy priorities of previous decades.

The Green Deal is a milestone in the history of the EU and beyond because it is the first time that an ecological transition has been the main policy priority of an economy of about 450 million people. It contrasts starkly with the disastrous austerity agenda pursued by the previous Juncker Commission, which effectively deprived national economies of the means to invest in the transition.

This outdated agenda rested on a consensus among the traditional parties that the European project revolved primarily around the single market, which would in turn enable economic growth. This market-driven integration of member states represented the main engine to achieve one of the foundational ideas behind the EU: namely, to keep the peace in a diverse and fractious continent.

This key concept developed simultaneously with the social contract adopted in Western European countries after World War II. To keep the peace between workers and capitalist employers in the context of the Cold War and the threat of social unrest, the social contract ensured that both groups were entitled by the state to share the spoils of economic growth through direct negotiations.

Such a contract works well in a context of economic expansion and global dominance. However, with the advent of geopolitical shifts and the tendency towards secular stagnation in richer countries, this deal has become outdated. The massive increases in wealth inequality and the stagnating incomes of the (lower) middle class in recent decades confirm the breach in the old social contract.

The old system is increasingly abandoning the least well-off. While a cosmopolitan and privileged group of people reaps the benefits of decades of hyperglobalization, many of the working class and those employed in industrial sectors that relocated outside of Europe have lost a sense of security, perspective and respect in society. No wonder many of them have taken to the streets in recent years.

Critically, we are now witnessing the wholesale environmental destruction of our planet, much of it fueled by economic growth. This makes it clear that the violation of the social contract cannot be solved by reverting to the old agreement. On the contrary, to avoid destroying our natural life-support systems, we will need to reconcile social justice with environmental and climate justice.

Therefore, we need a new deal: not only to keep the peace in Europe, but also to make peace with the planet. That is the challenge of the twenty-first century; it is also why the Green Deal should not simply be a new growth strategy, as the Commission claims, but should serve as a new and equitable social contract between society, the economy and our living planet.

Before concluding on this point, let us first dive deeper into the plans of the Commission, to see to what extent social fairness or justice is included.

THE JUST TRANSITION OF THE EUROPEAN COMMISSION

The 'Political Guidelines' of the Commission led by Ursula von der Leyen make it clear that the key priority of her team is to achieve climate neutrality by developing a Green Deal and that another of its priorities is to ensure a 'just transition for all'. As indicated above, one of the main levers of the Green Deal is to arm the EU with massive investment capabilities.

Soon after receiving its mandate, the Commission launched its Sustainable Europe Investment Plan in January 2020 to mobilize

€1 trillion in sustainable investments. As part of this plan, a newly set-up Just Transition Mechanism would marshal at least €100 billion in investments over the period 2021–2027 to support workers and citizens in the regions most impacted by the transition.

Part of the fund would come from the EU budget, while the rest would be co-financed by member states, complemented by contributions from InvestEU (an investment programme aimed at long-term funding for businesses) and the European Investment Bank (EIB). Extrapolated over 10 years, the Just Transition Mechanism would amount to around €143 billion in funding.

However, the Covid-19 pandemic hit Europe just a few months later, prompting the Commission to launch its NextGenerationEU recovery plan at the end of May 2020. The plan doubled down on the investment agenda already set out in the Green Deal. This also led to a major overhaul and strengthening of the Just Transition Fund, which is the first pillar of the Just Transition Mechanism.

The budget of this brand new fund has been more than quintupled, from €7.5 billion to €40 billion. A mere €10 billion will be drawn from the EU's regular long-term budget for 2021 and 2017. The EU Covid-19 recovery fund covers the remaining €30 billion. This is a little-known element of the funding structure, but it is worth mentioning because it means an increase in intra-EU solidarity.

The reason why it promotes solidarity within the EU is that about half of the NextGenerationEU agreement is funded by the collective issuance of bonds at EU level. Raising debt collectively in order to support climate and social transition policies through the Just Transition Fund means that EU member states effectively help the regions that are most affected by the shutdown of highly polluting industries.

Plant closures are primarily occurring in eastern Germany and in Central and Eastern Europe (Poland, Romania, the Czech Republic and Bulgaria), but they are also found in Southern Europe (France, Italy, Spain, Greece, etc.). According to the initial proposals, Poland would get the largest slice of the fund, followed by Germany. It remains to be seen how the final distribution of funding will take place in the coming years.

The Just Transition Fund alone will generate between €160 billion and €260 billion in investments thanks to additional national co-financing and the use of existing EU cohesion funds. The Commission also strengthened the other pillars of the Just Transition Mechanism. Infrastructure investment through InvestEU went up from €10 billion to €20 billion, leveraging €90 billion in final investments.

We can add to these impressive numbers another €25–30 billion from the Public Sector Lending Facility of the EIB, which partly benefits from EU funding, aimed at financing local public authorities. This money will be used for social support such as (re)training, income support and other employment policies, but also for the deployment of new industries and land restoration.

These new sources of funding may sound impressive, but the idea is not new. Back in 1951, the European Coal and Steel Community created a 'Fund for the training and redeployment of workers'; this later led to the European Social Fund. which today comprises about 10% of the total EU budget. However, as free market liberalism increased its dominance, the idea of just transition languished.

We owe the renewed interest in this area and the introduction of just transition funding in the Green Deal to the international labour movement. By introducing it into the heart of the International Labour Organization as well as entrenching it in the 2015 Paris Climate Agreement, it found its way into the EU's Energy Union and a newly created Platform for Coal Regions in Transition in 2017.

Following its inclusion in the Paris Climate Agreement, the European Parliament proposed a Just Transition Fund in 2016 and again in 2018, in the context of the Multiannual Financial Framework for 2021–2027. The Parliament proposed a fund of just €4.8 billion. This is peanuts compared with the €40 billion finally set aside in May 2020. The impact of the Green Deal is clear for everyone to see.

IS THE EU DOING ENOUGH?

The Just Transition Mechanism sends out a clear political signal that political leaders want to help citizens who are likely to lose their jobs. This creates a double dividend for member states: it creates a basic sense of social acceptance among the population to combat the climate crisis, but it also lays the foundations of a new industrial ecosystem, providing locally anchored quality jobs.

Member states have to draft territorial just transition plans to access these funds. This will enhance transparency and ensure that the citizens and territories that make the greatest efforts are the ones that receive the greatest support in return. But will this be enough? It is true that member states will receive according to their needs, but at the same time, these needs are much greater than the money that has been put at their disposal.

Having said this, we can extend this criticism to many realms of EU policymaking, of course. In that sense, the glass is actually half full rather than half empty, because it is the first time that the EU has actively tackled the challenges of economic conversion, social compensation and reskilling, as well as land restoration, in a coherent policy package with unprecedented levels of funding. We should take heart from this significant step forwards.

However, there is a broader concern when we consider the definition of just transition. The official just transition policies in the EU tackle the transition challenges and impacts for workers, particularly different geographical effects. Yet, these are not the only groups that bear the effects of the ecological transition: other vulnerable groups, future generations and people outside the EU are also impacted.

Take the people belonging to the lowest income brackets, who generally do not have any capital to invest in renovations, heat pumps or expensive electric vehicles. In fact, many of them have trouble paying their current bills let alone investing in future developments. For instance, 8% of the EU population say that they were unable to keep their home adequately warm in 2020.

It is true that policies in many member states try to remedy these groups' lack of money. Direct subsidies for the purchase of electric vehicles or renovations are a popular option. However, according to recent studies on the use of energy subsidies in the EU, 65% of these funds go to homeowners who already possess enough financial means to renovate and insulate their homes themselves.

This means that the subsidies do not lead to additional reductions in climate emissions. Worse still, these subsidies significantly increase the value of homes, leading to higher wealth inequality. In short, most of the time we are witnessing a Matthew effect – the rich get richer and the poor get poorer – while cleaner alternatives remain unaffordable to the most vulnerable households.

But there is more. The recent 2022 World Inequality Report shows that the poorest half of the populations of rich countries have already reached their 2030 emission reduction targets. Compare this with the richest 10% in a country like France: they emit five times as much per capita as the bottom 50%, whose incomes and wealth have stagnated. This is pretty galling, knowing that the rich have indeed grown even richer.

We should all contribute our fair share to make the ecological transition a success, but it is clear that the rich are not pulling their weight. It comes as no surprise then that a backlash from the public emerges in such conditions. In these situations, many people and policymakers cannot resist the urge to point at the '*gilets jaunes*' as proof that we should hold back on our climate ambitions.

The *gilets jaunes* are a good example, but not for the reason that most pundits assume. This becomes clear when we actually listen to their grievances. The protesters were against not environmental policies but rather their unfairness and the negative effects they have on poor people. For example, raising an ecotax in rural areas that were increasingly deprived of public transport was indeed a socially regressive policy.

These protests were instead a feature of the failed neoliberal policies of the last few decades. Raising the retirement age is likely to trigger a new round of protests. In short, all this underlines the importance of broader social justice, where social policy and fiscal policy have equally important roles to play in the Green Deal and the ecological transition in general.

THE IMPORTANCE OF MEMBER STATES

Our environmental and climate policy thus far has only harvested the low-hanging fruit. We have nearly exhausted the easy options; now comes the hard part. In the realm of climate policy, we will need to massively ramp up our annual emission reductions in the coming decade. This implies systemic changes and radical choices. The big question will be how costs and benefits are distributed.

There are two major reasons for considering distribution. The first is that the promises of the old social contract have eroded. Young people in particular do not believe they will be able to enjoy the benefits of ever-increasing consumption of goods and services, stable and long-term employment, and an adequate pension. Nor do 77% of the 15–35-year-olds surveyed in 23 European countries think our consumer habits are sustainable.

This means people are no longer looking at increasing the pie – they are instead looking at how the pie can be divided more equitably. This relates to the second reason for renewed attention to distributional effects, which is that most climate policies are regressive in nature. Low-income households generally bear more of the costs and impacts of both climate policies and the effects of the climate crisis.

Compare subsidies for rooftop solar panels or electric vehicles with programmes oriented towards renovating social housing projects. Budgets for the former are consistently higher than for the latter. Alternatively, compare the lack of carbon pricing and the many tax advantages bestowed on the aviation sector, which is disproportionately used by high-income households, with the lack of investment in rail transport.

We can avoid many of these adverse distributional effects by changing fiscal and social policy, but also by redesigning climate policies directly. This will be necessary, because the EU is a powerhouse when it comes to its internal market and its environmental policy, but it is a stunted giant when it comes to social and fiscal policy, which remains under the control of member states.

This means that the EU has to perform a complicated dance in designing its climate policies while taking into account the appetite for fiscal reform at member-state level. Let us illustrate this with a concrete example. In its climate package released in July 2021, the Commission proposed to extend the EU Emissions Trading System (ETS) to two new sectors: transport and buildings.

Expanding the ETS means that a system of carbon pricing will come into force in 2026, where the suppliers of transport and heating fuels will have to pay a price for climate emissions. These suppliers are major companies possessing the market power to transfer this extra cost to consumers, including small and medium-sized enterprises

(SMEs) and households. In the absence of complementary measures, poor households will be affected the most.

For instance, the 10% of families with the lowest income in the EU spend almost 10% of that income on energy expenditure, without even including transportation costs. Up to 90 million EU citizens (20%) face difficulty when it comes to accessing public transport. A price of €100 per tonne of CO_2 would increase road transportation and heating bills in the EU by around 25%.

We should keep in mind that poor households generally do not reduce their consumption when faced with a price hike in these sectors. If alternatives are not available or are unaffordable due to a lack of credit, poor households are simply forced to pay the higher price. If authorities reduce investment in public transport in rural areas, cash-strapped people have no choice but to resort to cheap and polluting petrol cars.

It is also important to note that the impact of a CO_2 price on transport and buildings differs depending on the member state. For example, a CO_2 price of €100 per tonne would lead to a 52% increase in spending on heating for the poorest 20% of households in Poland. This is twice the EU average. Geography has an influence on this, but so does the level of inequality.

Recent findings by the Stockholm School of Economics indicate that in high-income countries with relatively high levels of inequality, carbon taxation is regressive for necessities such as energy products. However, it is much more proportional in countries with low levels of income inequality. This raises two relevant points.

Firstly, it comes as no surprise that countries such as Finland, Sweden, Denmark and Norway all implemented carbon taxes around 1990–1992, a time when levels of inequality were relatively and historically low. Secondly, it proves that member states can offset the regressive effects of carbon pricing by reducing inequalities across society.

This can be done in two different ways. It can be achieved by ensuring adequate minimum wages, by applying much more progressive taxation across the board, and by expanding universal public services, particularly in areas currently deprived of them. An alternative method is to return revenues to households. Reducing taxes on labour income is one option, while distributing lump-sum transfers and investing in deep renovations for the poor are two others.

This underlines the importance of social and fiscal policies in member states. It is a matter of political choices and priorities. Politicians and parties pulling on the brakes when it comes to climate ambitions by invoking the argument of affordability and the impact on the poor are effectively abusing the plight of our most vulnerable fellow citizens in order to resist a fairer and more equitable society.

INCREASED EFFORTS AT EU LEVEL

While its powers in social and fiscal policy are limited, the European Commission does acknowledge that the EU could do more to mitigate the negative distributive effects of the ETS extension to the transport and building sectors. Fairly late in the process, it proposed a Social Climate Fund (SCF) in the form of a new regulation to address these social impacts.

The SCF was originally not part of the EU's climate package and therefore did not appear in the Commission's 2021 work programme. The proposal only appeared because of widespread criticism that the Commission failed to provide adequate support for more vulnerable consumers who could face the greatest difficulties in managing high energy costs.

The SCF has two purposes: to finance temporary direct income support for vulnerable households, and to support measures and investments that reduce emissions in the road transport and building sectors, thereby in turn reducing costs for vulnerable households and SMEs. Needless to say, setting up a compensation mechanism in a heterogeneous economic area with twenty-seven members is not straightforward.

In Bulgaria, 30% of households are not able to keep their homes adequately warm, whereas in Finland the equivalent figure is only 1.8%. By redistributing funds to the poorest households in the EU, the fund would become an instrument of solidarity across member states. Without a doubt, this is essential to convince member states with lower aggregate income to agree to more ambitious climate policies.

In total, the SCF will receive €72.2 billion of EU funding between 2025 and 2032. This sounds like a lot of money, but it represents only 25% of revenues raised by extending the ETS to transport and buildings. In order to use these funds, member states need to draw up social climate plans as well as co-financing half of the measures and investments mentioned in them. These member state contributions would effectively double the amount of funding going to social climate policies.

Social climate plans are a good idea. They ensure that national and EU spending priorities are aligned and complementary. But withholding a big part of the new revenues from the SCF is problematic. The purpose of a carbon price is not to raise revenue but to support people with switching transport modes and renovating their homes. In short, there is a strong argument to redistribute 100% of the funds.

If a large part of the revenue from the extension of the ETS to buildings and transport ends up in the general EU budget, the social acceptability of the ETS, and of climate policy in general, is at risk. It goes against all the evidence we have, which indicates that the income from carbon pricing should be visibly earmarked for a green and just transition.

Another issue is whether the SCF is big enough to make a real difference. Let us take the example of the Flanders region, where the cost for home renovations for vulnerable households amounts to anything between €1.4 billion and €6 billion per year. Initial calculations indicate that Belgium as a country would receive only €231 million per year from the SCF.

The objective is both to compensate for the initial price hike and to provide funding for new investments in deep renovations and transport infrastructure. It is clear that the SCF will not be able to deliver. Moreover, it will only come into force in 2025 – just one year before the carbon price is extended. As investments take years to materialize, the SCF will arrive on the scene too late.

There are also concerns about whether the funds will actually reach all the vulnerable households that need it. The proposal merely targets those households in energy poverty affected by higher ETS prices. This is too vague. It should instead specify households within the lowest income bracket, living in the worst-performing houses and/or with little alternative to individual car use for their mobility needs.

Does all of this mean that the SCF is already dead in the water and that we should abandon it? No, far from it. There is a good case to be made for such a fund, as the climate transition will inevitably have an impact on the most vulnerable in society. Until now, there was no dedicated climate law that explicitly addressed their plight. We should hold on to the concept and improve it.

We can beef up the fund with additional income streams, whether they come from the existing ETS or from other sources such as the financial transaction tax, a proposal that was already made by the Commission back in 2011 and that still to this day has not come into force. Furthermore, we should address inequity in all climate policies, e.g. by phasing out free emission allowances in the ETS and also the indirect-cost compensation for heavy industry at the expense of tax-paying consumers and governments.

Lastly, we have to accept the limits of the EU and its competences at this moment in time. Social and fiscal policies at member state level therefore need better alignment with the broader European climate agenda. That is why the Green Deal should be more than a technocratic policy agenda: it should in fact become a priority in all member states.

A GREEN DEAL AS A NEW SOCIAL CONTRACT IN EVERY EU MEMBER STATE

Techno-optimists insist that the right policy levers will enable technologies and systems to solve the climate and biodiversity crises. It is true that we still face real challenges in terms of planning, technology development and the roll-out of infrastructure for the ecological transition. But those optimists fail to pay equal attention to political consent – particularly the consent of those most affected by these policies.

The ecological transition, and by extension the Green Deal, will require us to make political choices leading to sacrifices for some and opportunities for others. We cannot sweep these difficult decisions under the carpet. If we do, populist extremist parties and move-ments will pop up everywhere to harvest the existential feelings of discontent among the disaffected.

It is all too easy for enemies of the Green Deal to point at rising energy prices and claim that ordinary people are sacrificed for the sake of the ecological transition. This is a cynical discursive inversion of the fact that the most vulnerable people actually suffer most from our current fossil-fuelled, linear and extractive economy. We should do everything we can to prevent that.

In addition to our moral duty to champion a just transition, this constitutes the strategic reason why the Green Deal absolutely needs to marry social justice with climate and environmental justice. It is delusional to think the Green Deal can succeed without doing so, even though, regrettably, some pundits and policymakers label this agenda as a 'leftist' plot to overthrow market capitalism as we know it.

That critique is wide of the mark. In the end, the question is not so much which technologies we need, but rather how much inequality and turbulence people accept for the sake of (ecological) progress when applying these technologies and the policies needed to realize them. Given the growing inequalities of today, the most likely answer is: not much.

What we need, therefore, is a radical overhaul of the old social contract. In the age of the Green Deal, the new social contract implies a better, healthier and more equitable life for people everywhere and greater care for the natural environment. We should move from a world of eternal scarcity and endless wants to a society providing universal basic services and abundance in the context of planetary boundaries.

Let us conclude with some practical suggestions on how to move forward with the Green Deal.

Firstly, the lack of public commitment to the Green Deal and its concomitant monitoring mechanisms from member states is a big problem. We need political leaders and mass movements to link Green Deal objectives with national, regional or even local issues. There has never been a better occasion for this. For example, iso-lated struggles in far-flung places have the potential to connect to other causes. In recent years we have already witnessed a number of good examples, including the 'Fridays for Future' movement and the many court cases against government inaction regarding the climate crisis. There is great potential to use the Green Deal to ramp up and connect such initiatives across member states.

Secondly, governments in member states should have a better understanding of the advantages of increased solidarity. The SCF provides a good example. Frugal member states do not want to extend solidarity to member states with lower incomes, and they prefer to keep all of the revenues generated in their home country from the new carbon pricing in the heating and building sectors.

However, the carbon price is determined on a European level. This means that poorer member states will face much larger price hikes, relatively speaking, with less income from the ETS to compensate the most vulnerable households. Governments in these countries will vehemently oppose more ambitious climate policies that the frugal, richer member states would actually support.

The same applies to the often-ignored topic of borrowing and issuing debt at EU level. We have already referred to the size of the EU recovery plan and how instrumental it was in beefing up the Just Transition Fund. The truth is that we will have to invest a lot more in new and collective infrastructure. Therefore, the Green Deal and just transition are directly linked to macroeconomic policy.

In the context of the pandemic, member states were ready to show solidarity by suspending EU debt rules and issuing common EU debt instead of leaving this up to individual member states. When the pandemic subsides, will we return to the old paradigm of fiscal consolidation? Will we raise taxes to pay back this debt at a time when we need citizens to invest in heat pumps, solar panels and insulation?

Could we instead create a structural investment capacity through common EU debt? Or perhaps we could remove climate investments when accounting for debt in each member state – a softer approach that may appease financial conservatives and progressives alike? The inconvenient truth is that these decisions will be more important for the just transition than most other political decisions. This is because they raise questions over who will pay for the transition, the degree of solidarity there is between richer and poorer member states, the volume of capital ready to be invested, the priorities of these investments, and much more. This brings us to the third and last practical suggestion, which is to open up the debate to citizens. Discussions on these topics usually take place among elites. That should change.

Practitioners of just transition keep repeating that social acceptability absolutely requires a dialogue with the people affected by the changes. When people feel deprived of any influence over or say on the matter, they become easy prey for populists. Policymakers and movements who want to transfer the Green Deal to the national or local level need to support citizen voices actively and forcefully.

There are many ways to organize this, including citizen assemblies, consulting or co-deciding with civil society, representing affected groups, and applying criteria of justice and equity in fiscal reform and other policies. The important thing is to acknowledge that governments and politicians no longer have the authority or capacity to impose big changes in society. We will have to do it together.

PART IV
Frontrunners Showing the Way

Just transition – youth perspectives

Sean Currie

If there is to be a just transition, it will not come easily. It will be the gradual result of an arduous political, economic and social struggle against vested interests and a now-dominant capitalist ideology.

From looking at the development of this struggle in Europe over the last 3 years, two things are clear. Firstly, it is winnable. Six million school strikers taking to the streets in one week, fifteen national governments in Europe declaring a climate emergency, and the EU pledging to be 'net zero by 2050' were all completely unimaginable prior to 2019. So who knows what we can do in the next five years? Secondly, young people will play a leading role in this struggle: young people of various demographics, different degrees of radicalism and with a wide range of carefully chosen strategies aimed at dismantling systems of exploitation. This dynamic – young people being among the leaders of the fight for a just transition – is what this chapter will explore.

In terms of understanding 'just transition', I follow the Climate Justice Alliance in using a broad definition that not only concerns workers, but also must 'transition whole communities' (Climate Justice Alliance n.d.). If we forgo this intersectional approach, I argue, then we risk leaving the door open to false solutions that leave in place oppressive power structures, such as patriarchy.

As a logical consequence of this global definition, the concept is intimately connected to that of climate justice, and it is with this concept that I will therefore begin the chapter. The need for a just transition movement will then be demonstrated through describing the failure of the European Commission in this area. Afterwards, we will look at how young people are bringing us closer to a just transition by analysing various parts of the movement, from an illegal occupation in Georgia to a corporate lobbyist in Germany, while not forgetting the usual suspects: Fridays for Future (FFF) and Extinction Rebellion (XR). I will argue that the strength of the movement lies in its strategic diversity, while powerful actors are nevertheless responding to it by undermining its radical, contested nature and depoliticizing its motives.

Throughout the chapter, I will argue that young people are reshaping the climate movement and thus their own generation's role in society – or, if I am to be more self-absorbed, I argue that *we* are reshaping *our* generation's role in society. As a 24-year-old activist-cum-political scientist, this chapter has been informed by my own experiences as much as by academic research. Importantly, it is also informed by interviews with diverse activists from across Europe and beyond, only some of which are included.

CLIMATE INJUSTICE

If you are reading this book, you would probably say that you support climate justice. But what does it actually mean?

Climate justice means recognizing that climate change is a scar of colonialism

Europe has emitted, and continues to emit, a vastly disproportionate share of greenhouse gas emissions (Rocha *et al.* 2015). Climate breakdown is undoubtedly causing destruction in Europe, but it is hitting communities in the Global South harder, faster and earlier. Not only does geographical bad luck make certain countries more prone to flooding, drought, extreme weather, forest fires and the spread of disease, but these countries have been made more vulnerable to their effects. A history of colonial exploitation and resource extraction has left a Global North–Global South divide in wealth and infrastructure that makes those countries that are most impacted by climate change less able to cope with its effects. It is in this context that a report was released in 2020 warning that up to 1.2 billion people could be made refugees due to climate change by the year 2050 (Institute for Economics & Peace 2020). For context, at the height of the so-called refugee crisis in 2015, just over 1.3 million people sought asylum in Europe (Pew Research Center 2016).

Climate justice therefore means centring the voices of indigenous people

One person I interviewed for this chapter was Amanda Luna, an indigenous woman born in Peru Huánuco, Quechua, who is a self-described 'activist for the defence of indigenous peoples' rights, and a defender of Mother Nature'. Raised in the Anda Amazonia region of Peru, Amanda has been living in Europe for 8 years. Having moved from South America to Europe, she says that her relationship with nature has changed, and therefeore her fight for the environment has changed too. 'In Europe, the power of nature is not perceived as strongly,' she says. Without trivializing the destruction that has been caused by the ferocious flooding in Italy, the deadly heat waves in France, the nightmare forest fires in Portugal, and so on, it is clear that we are more removed from the effects of climate change. In Peru, the infamous effects of the El Niño and La Niña weather patterns are becoming increasingly severe: not only in terms of direct destruction – through, for example, deadly landslides – but also in terms of the ever-increasing spread of diseases like cholera, malaria, dysentery and yellow fever. Historically concentrated in particular areas, these diseases are becoming increasingly widespread, something that Amanda attributes to El Niño and La Niña. Since this crisis stems from European colonialism, those who have benefitted from the exorbitant wealth bestowed by this exploitation should not seek to play a role in tackling climate injustice without listening to the people who have been exploited by it. As Amanda articulates: 'Climate change is a problem that sparks from colonialism... It starts from human beings going from a part of nature to wanting to dominate it.'

Climate justice does not, however, mean essentializing or romanticizing indigenous people

In Peru alone, there are sixty-five different indigenous communities, who speak more than ninety different languages. We need to be extremely careful, therefore, not to reduce this diversity when we highlight the voices of indigenous people. Amanda, for example, is speaking only from the perspective of Amanda, and never on behalf of indigenous communities – something she is very clear in emphasizing. Moreover, her perspective has been shaped from her experience as a Quechua person. The Quechua people have migrated a great deal in the last few decades, both within Peru and to Ecuador, Bolivia, Argentina, Chile and Colombia. Amanda's community, in particular, is therefore far from the essentialized image that many of us hold in Europe of indigenous people as environmental defenders with a deep historic connection to the particular land they live on. To quote Amanda:

> We often have a romanticised view of indigenous communities and how they protect the land, but there can be almost an irony in how they can be very dirty communities, plastic usage is a big problem… [Also,] not all indigenous communities have a right to save the land, there's much diversity. Some communities see the profit of receiving foreigners, and indigenous communities are also selling land to profit from mineral extractions.

Climate justice nevertheless means recognizing the innate injustice of extractivism

In order to satisfy the Global North's gluttonous consumption, multinational conglomerates have ravaged much of the Global South's resources. They have decimated essential sources of life, such as the Niger Delta, which has been turned into a black soup of oil, devoid of life, leading to the deaths of 16,000 babies within their first year of life in 2012 alone (Bruederle and Hodler 2019). Amanda's home city is the self-proclaimed 'mining capital of Peru', where extraction of gold and silver has contaminated the water, causing suffering that is unimaginable to most of us in Europe. The dangerous state of the drinking water combined with ever-increasing water shortages have caused an unsolved crisis. At least 2,000 children in the region live with chronic heavy-metal poisoning, while Amanda bemoans the regular diagnoses of leukaemia and anaemia among children. Extractivism, in short, treats not only the planet as disposable, but also frontline communities.

Climate justice means tackling climate injustice within Europe

Within our continent, the effects of climate change are felt most strongly by those who are least responsible. The EU's only recognized indigenous people, the Sami, face the destruction of their way of life as wildfires tear through grazing lands, as snow disappears and as the traditional knowledge of the local area is made increasingly obsolete by a changing climate and disappearing biodiversity (Şahin 2020). Meanwhile, working class communities and people of marginalized identities will be least able to cope with increased flooding, drought, heatwaves and all the economic and political consequences of climate change in Europe – something that has been exacerbated by decades of neoliberal politics that have dismantled support systems. Young people also suffer disproportionately from the effects of climate change, not only because of the time in which they will live, but also because many of them are more fearful about its present effects. In particular, they are already more likely to suffer from poor mental health, as an increasingly large number of young people are suffering from eco-anxiety (Gate-Eastley 2019).

Finally, climate justice means recognizing the injustice of pro-growth politics

Even without economic growth, we would need one-and-a-half planets to maintain current consumption, and 3% of economic growth per year means doubling the size of the economy in 23 years. This almost certainly means ever-increasing emissions, with no evidence for the 'green growth' ideology that is reliant on technology that does not yet exist – dangerously wishful thinking that the Intergovernmental Panel on Climate Change makes clear we do not have time for. Even if this technological utopianism were right, green growth promotes green extractivism, which simply shifts the target of exploitation. For example, rather than the 'soupification' of the Niger Delta, we will see repeats of the toxification of rivers in Tibet as a result of lithium mining. To quote Amanda: 'Colonialism has many forms today: extractivism, continuous economic growth … discrimination.' Our developed economies in Europe therefore need to move away from the damaging and unnecessary growth model and towards *actually* improving the lives of people in Europe and beyond.

The problem is that the European Commission has not got the message.

THE EUROPEAN COMMISSION'S FAILURE

'Europe's man on the moon moment' is how Ursula von der Leyen announced the European Green Deal. The speech adopted much of the rhetoric of the climate justice movement, claiming that 'we have to be sure that no one is left behind' and repeatedly talking about a 'just transition'. The same speech, however, made clear which economic system the Green Deal was made to protect, with the Commission's president asserting that 'the European Green Deal is our new growth strategy'.

So let us be clear: the European Green Deal will not tackle climate injustice. Undoubtedly, it is more ambitious about tackling climate change than anyone could have foreseen just 2 years before, and that is thanks to the incredible mobilization of the climate movement. However, any climate plan that is based on economic growth and the utopian technological dreams of 'net zero' targets is tantamount to a death wish.

Looking at the Commission's policy holistically puts into perspective their climate ambitions. The two big funding frameworks for member states – Next Generation EU and the Multiannual Financial Framework – require 27% and 30% of their funds, respectively, to be spent on a loosely defined concept of 'environmental protection', which is positive if underwhelming. These commitments, however, come at the same time as the European Commission ploughs ahead with the environmentally wreckless EU–Mercosur trade agreement, which will increase imports of beef and other land- and carbon-intensive products that are contributing to the destruction of the Amazon at a record rate (Lagoutte 2020). The trade agreement is so dangerous in the hands of Brazilian President Jair Bolsonaro that one indigenous leader has warned that 'signing this free-trade agreement could lead to genocide in Brazil' (Nelsen and von der Burchard 2019). This is just one high-profile example of how the European Commission is compartmentalizing climate change by setting aside a pot of money for 'protection' while simultaneously protecting economic growth at all costs – even at the cost of the lives of people in the Global South.

We need leaders who treat climate change as an issue of injustice and who tackle it holistically. To get there, we will need to fight for it.

THE CLIMATE MOVEMENT IN EUROPE

The climate movement in Europe is, really, a movement of movements. We will explore this by putting a magnifying glass up to a few small parts, thereby demonstrating its strategic diversity, which I argue is its greatest strength. Before analysing the usual suspects, we will start by looking at one local campaign in a part of Europe that is frequently forgotten in these discussions.

The campaign took place in Tbilisi, Georgia, and it successfully prevented a destructive commercial building project in the Dighomi Forest. The campaign was led by the Georgian Young Greens (GeYG), which was an independent youth organization campaigning on leftist environmental ideas.* I sat down for a chat with Giorgi Ptskialadze, an activist friend who was the co-spokesperson of GeYG at the time, who told me everything I know about the campaign.

The Dighomi Forest, or the so-called lungs of Tbilisi, is a publicly accessible 'forest park' that plays an important role for the local community, as well as for the city as a whole. The forest park, which has been cut back by 75% since the fall of communism, provides a rare green space for the population. There was dismay among the GeYG, therefore, when it was revealed in 2019 that a private company would build a 4,000-square-metre restaurant in the forest, plus related infrastructure. How the group organized to prevent

---

*The organization has since, impressively, begun the process of becoming a fully fledged political party with its own youth wing.

this destruction provides an interesting case study about the role of 'youth' in garnering support, the role of creativity in building transformative campaigns, and the role of local community campaigns in the global movement for climate justice.

Campaign strategy and tactical diversity

The GeYG's campaign went through multiple stages, and it mixed online and offline methods. They started with raising awareness: distributing homemade leaflets, knocking on doors and posting within local community Facebook groups, all to tell local residents about the planned destruction. They combined this with a digital campaign targeting the Tbilisi population more generally, posting videos of 2–3 minutes in length on both their organization's Facebook page and a specially created 'Save Dighomi Forest' page. The tone varied from serious conversations between older men on a couch, to comedy in the form of a parody promotional video that went viral. They then started posting a petition link under the videos. According to Giorgi, this 'has literally zero legal influence … but is a good way of mobilization and spreading information'.

Only after they had built a demonstrable city-wide support base, along with an engaged and aware local community, did they organize the first demonstration. As Giorgi tells it:

> Before the demonstration we said that we're going to camp there, so we stayed overnight. And then on the morning we had a demonstration, and it drew quite large numbers. Nobody expected that many people, and what's most interesting is that it was mostly local people who attended the demonstration.

Around 200 people flocked to the forest, symbolizing a huge show of strength by the campaign after months of work.

The diversity of offline and online tactics in the campaign stretched beyond even what the GeYG was doing. At the same time as they were engaging in social media messaging, community organizing and mobilization, a prominent non-governmental organization (NGO) called the Georgian Young Lawyers Association (GYLA) was battling in the courts to prevent the restaurant being built. Giorgi explains that these two parts of the movement complemented each other because, unlike in Western Europe, the court is not free from the influence of the government: 'You have to have a very strong movement behind the case, because otherwise the court will say what the government will say, and they should be kind of afraid of making such decisions [against the government].'

The situation came to a head in August 2019. With the local court repeatedly delaying a decision, the company started moving construction machinery to the proposed site. Expecting construction to start immediately, the activists moved to occupy the site. Between 100 and 200 people joined a short march to the site, at the end of which Giorgi expected to find police and locked gates protecting the private property: 'We even planned … who's going to be first arrested and if everyone is arrested who's going to be responsible for the continued continuation of the protest. Everything was planned. We expected that to happen.'

However, when they got there, they were shocked to find the gates open; nobody stopped them from going in. Giorgi and his co-spokesperson Mariam Vatsadze climbed on top of one of the machines and, in an act of poetic defiance, placed a small potted tree on top of the machine. Following varied speeches, the locals went home, but a small group of GeYG activists camped at the site for a few nights. When nothing happened, they packed up and left because, to quote Giorgi, 'there was no point' in staying. Just 3 days after the final activists left, the court ruled in favour of the activists.

Power, co-option and depoliticization

One of the most noteworthy things about the GeYG campaign was the way that it related to powerful actors. For example, at one point during the campaign, the mayor of Tbilisi, Kakha Kaladze, visited the site and took a selfie video, in which he talked up the positivity of the proposed restaurant. He praised the 'young people' protesting against the site as admirable and said that they were fighting for the right thing but were simply mistaken. The prevailing authenticity of the young activists was similarly shown when they respectfully gate-crashed a presentation by the company building the restaurant, who responded by thanking the young people for being critical and asking good questions.

This portrayal of young climate activists as admirable but naive is not unique to the Dighomi Forest campaign. In the early days of FFF, centrist and conservative politicians and media tried to ignore the movement. Then, when that became impossible, they belittled the movement's civil disobedience by focusing debates on 'skipping school'. Finally, they were persuaded to praise the movement, but have depoliticized it by characterizing the protestors as 'inspiring' school children who make up a broadly homogeneous and apolitical group – more akin to a group of angry teens than a highly political movement (Rucht 2019). Critics have alternatively downplayed the political and contested nature of the movement by reducing it to individuals: particularly Greta Thunberg, but also national figures such as Luisa Neubauer in Germany.

This is part of a broader depoliticization of the climate movement, which is most obvious with FFF. FFF is a diverse and highly political movement that defines itself as being for 'climate justice'. The political concerns and demands of the movement vary greatly both between and within countries (Marquardt 2020). There is debate about how radical reforms should be, how much focus should be given to individual behaviours, and how professionalized the movement should be. However, throughout FFF's messaging – from official statements to protest placards, from Greta's 'how dare you' speech to academic interviews with activists – there is a vision of a society that is different. The message quite clearly calls for a radical change from business-as-usual to something that respects planetary boundaries and that does not exploit marginalized people, either globally or within the Global North. This has led to intense debate within the movement, such as when the global FFF movement made a statement against Israel's 'settler-colonization', which FFF Germany condemned as 'antisemitic'.

Contrast this political movement with its portrayal by mainstream politicians and media. Jen Marquardt carried out a lengthy study of the movement that demonstrates that FFF's actions reflect 'fundamentally political struggles, but that politicians and media respond by framing the movement as apolitical, and the challenges of climate change as technological' (Marquardt 2020). This defence mechanism for business-as-usual is not new (see Swyngedouw 2011), but it is increasingly translating itself into vague, technology-focused climate legislation, such as the European Green Deal.

Another way that powerful actors relate to young climate activists is by feeding off our authenticity. World leaders from Ursula von der Leyen to Pope Francis have sought to bask in the legitimacy that FFF, and in particular Greta Thunberg, can give them. To give one bizarre example, Luisa Neubauer of FFF Germany was offered a seat on the board of Siemens Energy, who were under fire at the time for a planned coal mine in Australia (she rejected the offer).

This flattery is often even more sinister, seeking to destroy our authenticity by co-option. The GeYG case illustrates the difficult balancing act we must perform when seeking to influence leaders while trying to avoid co-option. The group was offered the opportunity to meet representatives from the Tbilisi City Council in person, but they insisted they would only do so if it was in public and there were witnesses. The City Council refused.

While this balancing act is a challenge, there is no doubt that the 'youth authenticity' is a powerful resource for activists. At 2019's 25th Council of the Parties in Madrid, for example, NGOs flocked to give FFF activists their badges, knowing that they could have more influence over proceedings. However, FFF has morphed from being a school strike to become a diverse intergenerational movement, with the average age of protest participants in different major cities ranging from 16 in Amsterdam to 40 in Brussels; the older adults are invariably there in solidarity (Wahlström *et al.* 2019). Even in groups like Scientists for Future and Parents for Future, adults are explicit in their aim of amplifying the messages of young people. This is, I argue, a fundamental repositioning of the role of young people in society. While young people have always been present in the story of environmentalism in Europe, prior to 2018 this was generally as passive victims who needed protection from the future consequences of adults' actions. Instead of being merely victims of climate change, young people are now the leaders of the climate movement.

The role of youth in age-inclusive movements

It is clear, however, that we still suffer from discrimination within movements. Occasionally, this manifests in discriminatory individuals who dismiss young people as 'naive', as 'rabble rousers', or as other clichés, but usually it is more subtle. One way it shows itself is the valuing of experience above all else and the pursuit of perfection. Another is typologizing young people. Many of us have sat in long, inaccessible meetings and have only been asked our opinion when it has related to things like social media or organizing parties. This is not simple to overcome, particularly because it is overwhelmingly performed by caring and well-meaning activists. But as with all forms of discrimination, we have to build our movements to not only include young people, but also to actively grow their power and experience. For me, this starts with prioritizing capacity building and the decentralization of power, and also with taking an approach that expects and accepts mistakes as part of the process. As young people within age-inclusive movements, we can best build power and push our movements to be more radical and inclusive by working as a caucus, as XR Youth have shown.

XR Youth is the youth wing of XR, which has been an enormously transformative movement, especially in the UK, but which has been criticised for a lack of racial diversity as well as exclusionary tactics and framing – particularly related to the glorification of arrests. XR Youth in the UK is, according to some of its own members, more diverse, inclusive and intersectional: more based on centring indigenous communities and voices from the Global South, and with more of a focus on 'climate justice than the main XR' (Gayle 2019).

A particular case in point was XR's planned action to shut down Heathrow Airport using drones. There were a number of concerns raised, including safety, public reaction and the fact that Heathrow is an airport that serves people from various socioeconomic backgrounds – in contrast to the business-elite-friendly London City Airport, for example. A lot of young people felt frozen out, with the Heathrow action presenting 'just the tip of the iceberg', so they decided to confront leaders in the organization. In July 2019, XR Youth disrupted an XR strategy meeting, armed with pastries emblazoned with the words 'we love you' on top. They made three demands, including that the Heathrow action be abandoned by XR and that XR Youth 'be supported in co-creating a new and healthier way forward for the organism' (Taylor 2020; Extinction Rebellion 2019). This show of strength worked: XR responded by cancelling their involvement in the Heathrow action and pledged that XR Youth would 'guide Extinction Rebellion UK through these challenges'.

BEYOND DISRUPTION

So far, the discussion has largely revolved around non-violent direct action – which I understand to be political activism that is both disruptive and conflictual without using violence or the threat of violence (Sharp 2005; Schock 2013). However, as well as disrupting institutions and powerful actors, young people are also fighting for climate justice within these institutions and with these actors. Side-stepping the myriad ways they are doing this through party political activism, I will focus on the ways that young people are using two arenas that have often acted as restraints on the power of activists: litigation and corporate lobbying.

Climate litigation: our newest weapon

While the criminal justice system is often a source of oppression for activists, particularly in non-democracies and particularly against the working class and people of colour, the increasingly powerful and increasingly widespread strategy of climate litigation is turning this on its head. This is a relatively new strategy, where individuals or organizations sue companies or public bodies (e.g. governments), either for their policies or for specific projects that contradict their climate obligations. The legal aims of these cases are varied, with some seeking to force climate action or to prevent destructive practices.

For example, in the case of *The People vs Arctic Oil* in Norway, the organizations Youth and Nature and Greenpeace argued that the granting of new drilling licences was unconstitutional. Some cases seek compensation, such as a 2019 UK Supreme Court case relating to the contamination of drinking water by the international mining

conglomerate Vedanta, which ruled in favour of Zambian community members. The case, which had legal standing because Vedanta
is headquartered in London, is particularly encouraging because it
creates a precedent for frontline communities in the Global South to
take litigation action in European courts.

Young people and people from frontline communities have an
especially important role in climate litigation. Again due to their
heightened legitimacy on the issue, as the victims of climate change,
these groups are frequently the plaintiffs* in cases. The paradox of
this is that it is a difficult strategy for these groups to engage with
because of its reliance on legal expertise. I heard from two young
women who have been involved in high-profile climate litigation
cases in Europe, and who greatly emphasized this point. First, there
is the problem of money. 'The best way to have a good case is to
pay for good lawyers,' one of them told me: to solve the issue of
money, litigation cases tend to rely on either *pro bono* lawyers or
crowdfunding to pay the fees. However, the activists still need to
overcome immense barriers of knowledge. Activists in litigation
often talk of the difficulty of listening to different lawyers who give
totally contradictory analyses of the legal basis of a case. The other
activist I spoke to, who is a Master's student in law, said: 'I still struggle not to be intimidated by lawyers.' Both activists spoke of often
being patronized by lawyers, who do not take them seriously. While
they both see this as age-based discrimination, these barriers are
undoubtedly higher for people of overlapping marginalized identities and for those from underprivileged socioeconomic backgrounds.
Relatedly, when young people and frontline community members
act as plaintiffs in climate litigation cases, there is a serious risk that
they are used as faces for the case without being awarded agency in
setting its direction.

A timely example is the ongoing Youth4Climate justice case, in
which six young people ranging from 6 to 22 years of age are taking
thirty-three European countries to the European Court of Human
Rights. They come from the Leiria region of Central Portugal that
forest fires tore through in 2017. Sixty-six people were killed in the

* That is. the person or organization who brings the case to court.

forest fires, most of whom were burnt alive as they tried to flee the fires by car. The case has used crowdfunding to overcome the financial challenges and is supported in expertise by the NGO Global Legal Action Network, which specializes in cross-border legal actions for justice against powerful actors.

As climate litigation becomes increasingly common in response to increasingly abundant climate agreements, there are other initiatives that seek to help it spread, such as the 'Leave it in the Ground' initiative's 'Climate Litigation Wiki'. It is worth noting that the hundreds of climate litigation cases in Europe have so far been mostly concentrated in Western Europe, Northern Europe and Spain, which can be explained by both the institutional factors and the strength of the climate movement in these regions. But as more cases set legal precedent for climate litigation, as climate agreements provide a stronger legal basis, and as the climate itself breaks down, this strategy is only likely to become more powerful and more widespread.

Climate lobbying: beating them at their own game

I first properly met Nils in Belgrade in 2019. We were taking part in a workshop on how to make graffiti stencils, planning to plaster them around the city during the night. He was wearing his iconic black hoodie with the yellow 'Refugees Welcome' logo emblazoned on the front. Between my amateurish attempts at scratching leftist slogans into cardboard, I spent much of the evening speaking to Nils about his experience with antifascist activism in Dresden and organizing with Ende Gelände. Nils would go on to become an executive committee member of the Federation of Young European Greens (FYEG), the independent youth wing of the European Green Party, in which he worked as a link between FYEG and social movements. This is all to say that Nils's activism has been long-running, diverse and oriented around social movements. So you can imagine my surprise when he told me he was going into corporate lobbying for a wind energy company in Germany. Motivated more by the climate emergency than his desire to sit in meetings with lobbyists from Shell, Nils's strategic turn offers a thought-provoking way of viewing activism, so I sat down to chat with him about how corporate lobbying can be a force for good.

Firstly, Nils argues that social movements are hindered by the power of fossil fuel lobbyists. Having been to every mass action by the movement since the summer of 2017, he is enormously positive about the work of Ende Gelände, but he says that he has 'also encountered the limits of it':

> We've done great at raising the topic and we, together with the Greens, forced the coal parties [CDU and SPD] to pass a coal exit law. But lobbyists have been so successful ... that the coal exit law is effectively a coal prolonging law that is giving fixed exit dates that are way later than the market-driven exit of these power stations, and they're getting 4.4 billion euro for it.

It was this observation that prompted him to make what he calls 'a strategy change', and he now plans to work in corporate lobbying for 2 years before reflecting on 'whether it's worth it'.

This strategic shift by Nils has been made with a view to strategic diversity within the climate movement:

> Every actor and every form of activism has a different role and a different strategy. For movements and for people active in youth politics, they need to talk another language than a lobbyist, because they're doing different things, though they can have the same goal. Like for me ... when I'm storming a mine and when I'm giving my input to a lobbyist paper of the German Energy Association, I'm trying to do the same thing. I'm just using totally different languages. And I think that we need to use all languages to really achieve the change that we want to see because when we only speak one language, the other languages and other parts of decision-making and power structures are just gonna break our neck.

Nils hopes that other activists will think about going into lobbying for the transition. I hope that corporate lobbying can be changed from something that is slowing down the energy transition a lot, and that is usually 'breaking our neck' as climate activists, into something that is neutral or even positive for the energy transition. However, similar to climate litigation, he is aware of the barriers to participation that exist:

It's also a question of privilege, because people who are active in social movements … nearly never have the chance to go into lobbyism because for lobbyism … it's much easier if you're a white male from an academic background, which are privileges that I have and that are making it relatively easy compared to others. I can just take off my black hoodie put on a suit, and shave, and then they're gonna think that I'm one of them although I'm not. But this privilege, only a small number of people have.

When I started drafting this chapter, I did not expect to be supporting a call for activists to go into corporate lobbying. As individuals there is a grave risk that we are corrupted by the system when we use this strategy – something Nils tries to prevent through the people he surrounds himself with and the activism he performs in his spare time. Nevertheless, the risk is there, and if the corporate world gets too crazy for him, he is open to another shift of strategy.

As a movement, there is an even greater problem. Despite the fact that NGO lobbying has long been an important strategy for environmental organizations such as Greenpeace, corporate lobbying is exclusively used by private companies to maximize their profit. It is therefore serving green capitalism, not climate justice, where the latter requires a disruption of exploitative profit motives, such as through energy cooperatives. However, I do not think this means we should not engage with corporate lobbying. Rather, it necessitates a diverse strategy. Like Ende Gelände's 'fingers' invading a coal mine from all angles, the climate justice movement needs to embrace strategic diversity. If we only did corporate lobbying, then we would never disrupt the existing economic order. But if we *only* take non-violent direct action, then we will let fossil fuel lobbyists control the rules of the game. Diversity is our strength, and we all have different roles to play in that.

CONCLUSION

In this chapter I have argued that young activists are playing a leading role in the fight for a just transition and are, out of necessity, employing a wide range of strategies. Some of these strategies, such as corporate lobbying, will be more focused on 'climate' than on 'justice', which necessitates an unrelenting pursuit of intersectional climate justice by other parts of the movement.

The question this raises is, if strategic diversity is central to the strength of the climate movement, how far are we willing to push it? In his provocative book *How to Blow up a Pipeline*, Andreas Malm argues that it encompasses violence. I remain unconvinced by Malm, but I believe that as a movement, we must create space for (sensitive) debate on these questions of strategy. In our social media age, it is easy to start sneering at other parts of the climate movement. It is easy to sneer at XR as privileged, out-of-touch 'pacifists' (which they are not, despite what Andreas Malm argues). It is easy to sneer at FFF for not being radical or political enough (I am looking at you, Tadzio Muller). It is easy to become tribal with our own sections of the climate movement, and to undermine the diversity of thought within other sections. What I have tried to demonstrate here is that this undermines the strength of the climate movement.

To win a just transition, we need to embrace strategic diversity, and we need to remain self-critical while focusing our anger outwards – at those who stand in the way of climate justice.

Bibliography

Bruederle, A., and Hodler, R. 2019. Effect of oil spills on infant mortality in Nigeria. *Proceedings of the National Academy of Sciences of the United States of America* 116(12), 5467–5471 (doi: 10.1073/pnas.1818303116).

Climate Justice Alliance. n.d. Just transition principles. Report, Climate Justice Alliance (https://bit.ly/3Cu2pee).

Extinction Rebellion. 2019. XR youth and Extinction Rebellion: forging a new relationship. Website, Extinction Rebellion, 24 July (https://bit.ly/3J1K1vy).

Gate-Eastley, J. 2019. Eco anxiety: climate change and youth mental health. *Green World*, 3 September (https://bit.ly/3HVeHxh).

Gayle, D. 2019. Does Extinction Rebellion have a race problem? *The Guardian*, 4 October (https://bit.ly/3hS3vqz).

Institute for Economics & Peace. 2020. Ecological threat register: understanding ecological threats, resilience and peace. Report, Institute for Economics & Peace, Sydney (http://visionofhumanity.org/reports).

Lagoutte, J. T. 2020. The EU–Mercosur trade deal must be stopped. *Green European Journal*, 15 May (https://bit.ly/3sUnziy).

Marquardt, J. 2020. Fridays for Future's disruptive potential: an inconvenient youth between moderate and radical ideas. *Frontiers in Communication* 5(48) (doi: 10.3389/fcomm.2020.00048).

Nelsen, A., and von der Burchard, H. 2019. EU's green trade promises face rainforest reality. *Politico*, 18 June (https://politi.co/34yoOdW).

Pew Research Center. 2016. Number of refugees to Europe surges to record 1.3 million in 2015. Report, 2 August, Pew Research Center, Washington, DC (https://pewrsr.ch/3Cu1U3Q).

Rocha, M., Krapp, M., Guetschow, J., Jeffrey, L., Hare, B., and Schaeffer, M. 2015. Historical responsibility for climate change: from countries emissions to contribution to temperature increase. Report, November, Climate Analytics, Berlin (https://bit.ly/3hR7OlV).

Rucht, D. 2019. Faszinosum Fridays for Future. *Aus Politik Zeitgeschichte*, 15 November (https://bit.ly/3sUe8zC) (In German).

Şahin, G. 2020. Peoples climate case – families and youth take the EU to court over its failure to address the climate crisis. In *Standing Up for a Sustainable World*, edited by C. Henry, J. Rockström and N. Stern, pp. 171–178. Cheltenham: Edward Elgar.

Schock, K. 2013. The practice and study of civil resistance. *Journal of Peace Research* 50(3), 277 (doi: 10.1177/0022343313476530).

Sharp, G. 2005. *Waging Nonviolent Struggle: 20th Century Practice and 21st Century Potential*, p. 10. Manchester, NH: Extending Horizons Books.

Swyngedouw, E. 2011. Depoliticized environments: the end of nature, climate change and the post-political condition. *Royal Institute of Philosophy Supplements* 69, 253–274 (doi: 10.1017/S1358246111000300).

Taylor, M. 2020. The evolution of Extinction Rebellion. *The Guardian*, 4 August (https://bit.ly/3gLFxwD).

Wahlström, M., Kocyba, P., De Vydt, M., and de Moor, J. (eds). 2019. Protest for a future: composition, mobilization and motives of the participants in Fridays For Future climate protests on 15 March, 2019 in 13 European cities. Report (https://bit.ly/3sWrDyN).

A just transition in agriculture

Anne Chapman

WHY AGRICULTURE NEEDS A JUST TRANSITION

The culture of agricultural communities is very different from that of post-industrial areas, but like the latter, agricultural communities have suffered from substantial declines in jobs over the past few decades and have lost much of what formerly held them together. The industrialization of farming, which has been happening since the 1950s but has gathered pace in recent decades, has involved many changes: mechanization; new, faster-growing but less resilient breeds of plants and animals; an arsenal of synthetic chemicals; and changes to farming practices, such as the move from sowing crops in spring to sowing them in autumn and from making hay to making silage.

Industrialization has produced cheap food but poor diets, has devastated wildlife, has polluted air and water and is a major contributor to climate change. It has also not been good for agricultural communities: the supply chain upstream and downstream of farmers has become consolidated and dominated by larger and larger companies, with whom farmers have little bargaining power. As a result, farmers have been trapped between the rising costs of their inputs and falling prices for their output, with the consequence being that their incomes have fallen in real terms, leading to the loss of farms and farming livelihoods. Much of the available work on the mega farms that have replaced small family farms is done by migrant labour, because working conditions are poor. There is a danger that those in rural communities who have lost out, feeling ignored by seemingly prosperous cities, turn to political extremists who at least seem to be able to give them people to blame for their plight: migrants and the 'metropolitan elite'. Thus, in 2016 the base of support for Trump was in rural America, and in the United

Kingdom support for Brexit was highest in Lincolnshire, an area of intensive arable agriculture.[*]

Agriculture needs to change to tackle the twin crises of climate change and biodiversity loss. Some think that the future will involve the further intensification of production, to produce more food on less land, perhaps using novel technologies that create plant-based or lab-grown meat, eliminating livestock and freeing land for nature. This future will enable large agrochemical companies, food manufacturers and retailers to continue to flourish. However, it would be at the cost of jobs and livelihoods in the countryside, and it would do little to improve biodiversity, the restoration of which needs to go hand in hand with reducing greenhouse gas emissions. This is because, as we have seen, increased production does not necessarily free land for nature: rather, we are over-producing a narrow range of arable crops, losing diversity in agricultural systems and in our diets. Overproduction means too much food goes to waste[†] and it fuels industrial meat production, where grains and soya are fed to housed livestock. Also, when land that has had a long history of agricultural use is suddenly abandoned, biodiversity is not necessarily increased: for example, the removal of grazing animals may simply allow vigorous grasses and bracken to thrive, which outcompete other plants, thereby reducing the biodiversity of grasslands.

This chapter will set out an alternative future for farming based on a combination of regenerative agriculture and what I have termed 'farming for nature'. These are not mutually exclusive practices: both may take place on the same farm, or even in the same field. They have been developed by farmers and landowners, and I will argue that they offer better prospects for more and better-quality work in agriculture than an industrialized agricultural system.

In industrialized agriculture the aim is to increase the yield of a particular crop or the growth rate of an animal. This is achieved by increasing inputs – of artificial fertilizers, pesticides, medication and animal feed – and the use of new, high-yielding or fast-growing (but

---

[*] See www.bbc.co.uk/news/uk-politics-36616028.
[†] Berners-Lee (2019, pp. 12–15) contains a good overview of how much food is wasted where.

generally less resilient) breeds of crops and livestock that depend on such inputs. In regenerative agriculture, on the other hand, farmers shift their focus from yield to profit margin per hectare. Increased profits are achieved through reducing inputs and improving the health of the soil and the wider ecosystem, so that nature does more for the farmer. Both ecosystem health and profits are achieved through increasing the diversity of crops, livestock, enterprises and also wildlife. In farming for nature, the focus shifts again, emphasizing the maintenance or restoration of natural habitats and processes, with food as a by-product.

This chapter is an abbreviated version of 'A just transition in agriculture', a report published in 2021 by the Green European Foundation with the support of Green House Think Tank. The report includes six case studies and some other sections (on the impacts of intensive agriculture, on the role of livestock in sustainable farming, and on the move away from maximizing food production) that have not been summarized here.

REGENERATIVE AGRICULTURE

Regenerative agriculture is an approach to farming that puts the health of the soil at the heart of the farming system. In many of its practices, regenerative agriculture is a form of agroecology.* However, agroecology has become a politically charged term associated with small-scale agriculture. With the exception of organic farmers, who often see what they do as a version of agroecology,† it is not a term used by many larger farmers who are nevertheless happy to say that they practise regenerative agriculture. In one way, regenerative agriculture can be seen as reincorporating some of the elements of mixed farming systems that declined as farms became more specialized, but in another way, it can be perceived as moving forwards to new farming methods, such as its use of cover crops, no-till arable systems and rotational or 'mob' grazing.

---

* See, for example, https://bit.ly/3he8Sjy.
† See https://bit.ly/3JPN6PA.

The four principles of regenerative agriculture are listed below.[*]

1. *Limit disturbance of the soil.* Disturbance, such as through plough-ing, destroys soil structure. In healthy soil, particles of clay, silt and sand are clumped together in aggregates that are held together by carbon-based glues produced by microorganisms in the soil. Pore spaces between the aggregates allow the infiltration of air and water. Ploughing disturbs this structure and lets lots of air into the soil, which accelerates the degradation of these glues by bacteria, releasing carbon dioxide. The growth of weeds may then be stimulated by the release of soluble nitrates from dead microorganisms. The destruction of soil aggregates reduces the porosity of the soil, making it more anaerobic and less able to hold water. Anaerobic conditions increase denitrification, in which nitrate in the soil is converted to the gas nitrogen – a pro-cess that also produces nitrous oxide, a powerful greenhouse gas. Ploughing also disturbs the network of mycorrhizal fungi, which play a critical role in enabling plants to derive nutrients from the soil. Soil is further disturbed by the addition of agrochemicals: nitrogenous fertilizers reduce mycorrhizal fungi because, in pro-viding nitrate to plants, they reduce the amount of surplus car-bon compounds that plants otherwise exude through their roots, feeding the mycorrhiza (Prescott *et al.* 2020); pesticides are toxic to various forms of soil biota, so they disturb the soil ecosystem.

2. *Keep the soil covered* through growing cover crops and through leaving crop residues on the soil. Ideally, there should be no bare ground at any time of year. Having living roots in the soil, all year round if possible, is important to feed the soil biology.

3. *Increase diversity* of both plant and animal species as much as possible. Diversity of plant species can be achieved by cover crop mixtures containing perhaps twelve plant species as well as increasing the variety of crops (together, as in companion plant-ing, or in rotations), animals and trees.

---

[*]These are taken from Abram (2020). Brown (2018) has five principles, with 'armour' (keeping the soil covered with crop residues) and 'living roots' listed sep-arately. Much of the discussion here is taken from Brown (2018).

4. *Integrate animals* into the agricultural system. Animals are always part of natural ecosystems, and grazing animals in particular have a key role to play in improving soil health.

For any farmer wishing to move away from conventional industrialized agriculture to a more regenerative system, implementing these principles is not easy: it takes time for the soil to improve, and compromises of some sort usually have to be made. Unlike organic agriculture, with its lists of approved and prescribed practices, regenerative agriculture is about a direction and a journey. One of its pioneers, the American Robert Rodale, defined it as 'a holistic approach to farming that encourages continuous innovation and improvement of environmental, social, and economic measures'.[*] Dan Burdett, who used his 2020 Nuffield Farming Scholarship to look at why farmers made the change to more holistic, regenerative practices, said:

> Of the farmers I met, 90% weren't organic. The majority of arable farmers still used some form of chemical input, preferring to keep all the tools at their disposal, but always looking to minimise their use over the medium to long-term. This makes [regenerative agriculture] accessible to all, and with no paperwork or inspection it is something that a farmer can start at any time and work out their own set of rules. This is in contrast to organic where the rules and regulations would certainly deter many farmers from making that transition.
>
> Burdett (2020)

Conventional arable farmers using no-till methods – where seeds are drilled directly into the residue of the previous crop, with no ploughing or cultivation of the soil – generally use a herbicide such as glyphosate (the only broad spectrum herbicide available to them in the EU) prior to planting a 'cash crop' (one they are growing to sell, as opposed to e.g. a cover crop). Many organic farmers, by contrast, find that they need to do some form of ploughing to control weeds

---

[*] See https://bit.ly/3pc9n1Q.

at some point in their crop rotation, though they may plough to a shallower depth than has become normal practice (Soil Association 2018). There is ongoing research into how no-till methods can work in an organic farming system,[*] and the report I referred to earlier includes a case study of one farmer who has grown heritage wheat on the same field for 6 years using a no-till organic system (Chapman 2020).

Grazing livestock play a key role in regenerative agriculture by helping to increase biodiversity and carbon in the soil. When animals consume plants they 'burn' some of the carbon in the plant to provide themselves with energy, ending up with more nitrogen than they need in their bodies, so they excrete nitrogen in the form of urea, thus returning it to the soil. Grazing animals also contribute the bacteria from their rumens to the soil. Animals (including those that live in the soil) thus increase the nitrogen–carbon ratio in the soil, enabling plants to grow and then extract carbon dioxide from the air through photosynthesis.[†] When grazing is managed in the right way it can therefore enable the soil to produce more than it otherwise would. Arable systems that do not use artificial nitrogenous fertilizers struggle to maintain soil fertility without including grazing livestock at some point in their rotation. Incorporating clover and grass leys (temporary pastures) into arable rotations, as is done in traditional mixed farming systems and in organic farming, builds up carbon in the soil, improves soil health and reduces the prevalence of pests and crop diseases.

Farmers practising regenerative agriculture generally use a form of rotational or 'mob' grazing: rather than being kept in a relatively large field for, say, a month, animals are confined to a small area using an electric fence[‡] and moved every day, not coming back to

---

[*] See https://bit.ly/35qQlOc.

[†] There is a good explanation of this on p. 11 of Chapman (2012).

[‡] Alternative systems have been developed that avoid the use of an electric fence: in the 'no-fence' system animals instead wear a GPS collar with a battery that gives them an electric shock if they go beyond the boundary the farmer sets on the system (accessed via a mobile phone). Before they get an electric shock the animals are played some music and, once trained, most turn around at this point. See www.nofence.no/en/.

that area until the grass has had time to recover and grow tall. Tall plants mean longer roots systems that can more effectively obtain water and nutrients from the soil. Productivity is increased because the amount of photosynthesizing a plant can do is proportional to the amount of green leaf it has – so a bigger plant makes more biomass each day than a small one. The aim in these sorts of grazing systems is for about 50% of the grass to be trampled down rather than eaten. Along with the dung and urine from the animals, the trampled plant matter feeds the biomass in the soil, adding to its carbon content.[*] Also, because the animals are moved away from their dung every day, their burden of parasites is reduced.[†] George Hosier, a farmer in Wiltshire who was interviewed for this project,[‡] says that since changing to this style of grazing, not only has he been able to stop applying artificial fertilizer to his pastures with no loss of grass growth, but he has also stopped his routine use of the medication that most farmers give to their cattle and sheep to treat parasites such as worms and fluke. This has increased the numbers of dung beetles (previously harmed by deworming medication) that incorporate the dung into the soil.

Non-grazing animals such as pigs and chickens can also play a role in regenerative farming systems. Gabe Brown, a North Dakotan farmer who has been a pioneer of regenerative agriculture, feeds grain screenings, which would otherwise be waste, to his chickens. In the summer his chickens are also put on pastures a few days after they have been grazed by cattle, in order to eat the fly larvae that have developed in the cow pats. He also keeps pigs, which in the spring use the pastures where his cattle have been feeding on bales of hay in the winter. The pigs stir up the residue of hay and manure, removing the need to harrow the land. In tree shelter belts the pigs root through old decaying wood, stimulating the growth of grass and herbs (Brown 2018, pp. 86, 89).

Gabe Brown tells the story of his journey into regenerative agriculture in his 2018 book *Dirt to Soil: One Family's Journey into*

---

[*] For more on mob grazing, see Chapman (2012).
[†] See https://bit.ly/3sgA0ot.
[‡] See https://bit.ly/3IB3dPI.

Regenerative Agriculture. He emphasizes that farmers should focus on profit per acre rather than on maximizing yield. Increasing profit is achieved not by using monocultures that rely on expensive inputs but by building up a diversity of enterprises that are synergistic with each other. He gives a 'cash flow' statement that shows how carbon (rather than money) flows through thirty 'products' for Brown's ranch. Gabe Brown and his family can manage all these different enterprises because they have radically cut down the amount of work they need to do:

> We don't have to haul and apply fertiliser, pesticides, and fungicides. We don't need to vaccinate and worm our livestock. We don't spend days chasing around the country to find the latest and greatest bulls, rams, and boars. We don't pregnancy test the cow herd, pigs, or sheep. We don't have daily chores of starting up farm equipment to haul feed to the livestock during the winter. We don't have to spend time hauling manure from the corrals out to spread on the fields.
>
> Brown (2018, p. 195)

A study of corn farmers in the North American Plains found that profits were best correlated with soil organic matter content, not with crop yield. Fields farmed regeneratively produced 29% less grain but gave 78% higher profits than conventionally farmed fields. Pests were also less abundant in regenerative farming systems without insecticides than in insecticide-treated corn fields (see LaCanne and Lundgren 2018).

The normal business assumption that has driven many farmers to try to increase their yields is that the way to increase profits is to increase output. Why this does not work in farming is explained in a report on hill farming in the United Kingdom, commissioned by a group of conservation organizations. The authors of the 'Less is more' report (Clark *et al.* 2019) examined the accounts of twenty-nine hill farms, as well as Farm Business Survey data of a further seventeen farms. For almost all the farms examined, if only farming activities were considered, the farms would be making a loss were it not for agricultural support payments; some were even making a loss with these payments. The cost of the farmer's own labour was not included in these calculations. However, the report reveals that farmers tend not to analyse their business accounts, instead looking only at total revenue against cost, so they do not realize that their farming activities are losing money (p. 12). The study found that many farmers assume that their variable costs are linear, increasing in proportion with their output. So if they increased their output sufficiently, they should get to a point where they start to make a profit. However, this is not the case. Up to a certain point, which the authors call the maximum sustainable output, farmers make use of free natural resources, of grass growth in the case of hill farms, and incur 'productive variable costs' – the essential/unavoidable costs linked to their farming system. To increase production above this point they have to buy in more resources – fertilizer, herbicides, additional feed, more medication, and so on. For UK hill farms (and for other types of farming that the authors examined) these costs are higher than the value of the additional output they make possible. That is, farmers would be better off financially if they produced less.

Farmers are often motivated to switch to more regenerative practices because of a desire to cut costs, though this can lead on to an interest in soil health and in making their land better for wildlife. Regenerative agriculture has clear benefits for the environment over the current industrial model and enables farmers to farm more profitably. However, at least initially, this switch is likely to reduce on-farm labour, as regenerative agriculture is about getting nature to do more of the work so that the farmer has less to do. Potential job losses may be countered by diversifying farm outputs, which means that more jobs can be supported. Diversification can add work at times of the year when there is otherwise little work, enabling full-time, year-round employment. For example, Whitehall Farm in Cambridgeshire has found that the integration of apple trees into its arable and vegetable-growing business (with the trees planted in rows to form windbreaks) has enabled them to employ someone full time all year round, because there is plenty of pruning and management of the trees to do over the winter. The farm can also add value to the apples themselves by processing them to make juice and by opening a farm shop to sell them directly to the public (Food, Farming and

Countryside Commission 2021, p. 40). A diversity of enterprises on a farm makes work on it more varied and more interesting, particularly given the continual experimentation, learning and innovation that characterize regenerative agriculture. It therefore has potential to provide high-quality, knowledge-based work.[*]

Reducing costs will mean farmers can retain more of the income they receive for their outputs, making it easier for farm businesses to make a profit and provide good-quality employment in agriculture. An example of what can be achieved is found at White Oak Pastures farm, which is owned by Will Harris, one of the pioneers of regenerative agriculture in the United States. The farm now employs 155 people in Bluffton, southern Georgia, regenerating what was a rapidly declining small town. Over the course of 20 years, Harris restored the degraded soil of his family farm through mob grazing with cattle, and he made enough money to buy up neighbouring land. White Oak Pastures now raises sheep, pigs, poultry, rabbits and cattle (ten different animal species in all), and it grows vegetables as well. It has its own on-farm slaughterhouse, which employs 120 of the 155 workers and which ensures that nothing goes to waste. Even the inedible viscera is used: it is composted for use on the vegetable fields (trying to find a use for the compost was what made Harris start growing vegetables). Along with employing so many people, White Oak Pastures has contributed to the regeneration of Bluffton through restoring the old Bluffton general store, which now sells general groceries as well as the farm's own products.[†] This is perhaps a model of what regenerative farming could do for rural areas in Europe.

FARMING FOR NATURE

In regenerative agriculture, the aim is to be profitable by reducing inputs and building up the natural capacity of the land to produce

---

[*] Greenham and Link (2020, pp. 49–50) argue that 'operatives' in industrialized agriculture become 'knowledge workers' in agroecological enterprises, whose know-how is essential in the experimentation, fine-tuning and learning processes that increase productivity in these farming systems.

[†] See www.whiteoakpastures.com and Burdett (2020, pp. 7–8).

good-quality food. Producing food is still key, but the idea that the yield of a crop or the growth rate of livestock should be maximized has been abandoned. In what I am calling 'farming for nature', on the other hand, the production of food no longer takes centre stage. Instead, food is a by-product of other activities whose main aim is to maintain or restore particular species, habitats or natural processes.

Though modern farming is highly destructive, abandoning farming in areas where it has had a long history does not necessarily improve things for the natural world. For example, many habitats such as species-rich limestone grassland or hay meadows become less biodiverse if farming ceases. The decline in grazing by goats and sheep in Mediterranean countries such as Greece has resulted in an increased risk of wildfires (Colantoni *et al.* 2020) and a decline in biodiversity.[*] In many places in Britain, too, the cessation of cattle grazing and of the practice of cutting bracken to use as winter bedding for livestock results in a monoculture of bracken, a fern that is not eaten by any grazing animals,[†] not the regeneration of woodland that some might expect. Trees cannot establish because they are shaded out by the bracken and are eaten by wild deer. Reintroducing grazing by the right sort of cattle can keep the bracken in check because the cattle (unlike sheep) are big enough to trample it; in other places, their hooves create pockets of bare ground in dense swards of grass where trees can become established.

The Burren on the West Coast of Ireland is an area with rocky uplands of limestone pavements that have particularly rich flora of wild flowers and associated insect life. It has been used for grazing as part of extensive farming systems for 6,000 years, but by the 1990s much of this had ceased as farmers had either left or switched to rearing fast-growing breeds of cattle on the lowlands, treating their fields with artificial fertilizers to maximize grass growth and making silage. This resulted in the pollution of water courses by fertilizer and slurry, and a decline in the biodiversity of the upland limestone areas.

---

[*] See references cited in Hadjigeorgiou (2011).

[†] Bracken contain several toxic compounds, though wild boar will eat the rhizomes (see https://en.wikipedia.org/wiki/Bracken and https://bit.ly/3IlZI0o). The lack of wild boar in most of the United Kingdom may be one reason why bracken is such a problem.

This decline has been reversed by a 'payment by results' system that uses EU funds to pay farmers for species-rich fields and clean water. In response, farmers have reinstated the practice of 'winterage', where cattle graze the rocky uplands in the winter, eating the tough, hardy grasses so there is more space for the rare flowers that appear in the spring. Critical to this system has been a change of mindset on the part of farmers, away from seeing what they are doing as just about producing food. One local farmer, Michael Davoren, is quoted as saying: 'In the past, the environment was a by-product. In the future, the environment is what we'll be producing, and the food will be a by-product.'[*]

Similarly, the cattle and sheep owned by Cath and Bill Grayson, who run the Morecambe Bay Conservation Grazing Company in North West England (see Chapman 2020, case study 5), are primarily raised to be conservation volunteers: they do produce meat, but this is a by-product of their role in maintaining species-rich habitats through their grazing. The cattle range over a wide area and can choose the grasses, herbs and tree leaves that they want to eat. Their grazing keeps in check vigorous grasses that would otherwise out-compete wild flowers, and it also tramples down bracken or rushes that would otherwise take over, checking the succession of species-rich grassland to scrub and then to woodland. On areas of rough grassland, for example, cattle often graze the herb-rich coverings of anthills that otherwise become overwhelmed by grasses, killing the ant colonies.[†]

Having a wide variety of plants available to eat, including the leaves of trees and shrubs, means that grazing animals are able to find more of the nutrients they need and to self-medicate when they are ill. This results in healthier animals, who are less likely to require veterinary medication, and whose meat and milk provide more nutrients for the people who consume them (Provenza *et al.* 2015). However, the ability to graze difficult ground and find sufficient nutrients is something that animals have to learn. It also takes a while for their digestive systems to develop so that they can cope

---

[*] See https://bit.ly/3shXyZW.
[†] From a personal communication with Bill Grayson, 2020.

with the lower-quality herbage that may be all that is available in winter.* For this reason, and because animals grow more slowly on less fertile land, conservation graziers such as Bill and Cath Grayson keep their animals for longer than the 36 months that is considered the upper limit for prime cattle. This can result in the meat not getting the top price you might expect: it is less tender than that of younger animals and needs slow and careful cooking, so is not favoured by supermarkets.

Farming practices can be changed to benefit nature on arable farms as well. This has been demonstrated by the conservation organization the Royal Society for the Protection of Birds, who in 2000 bought Hope Farm in east Cambridgeshire, a predominantly arable area, to trial and demonstrate nature-friendly methods of farming. In contrast to the ongoing national decline in farmland birds, Hope Farm saw a 226% increase in birds and a 213% increase in the number of butterflies by 2017.[†] This has been achieved by, among other things, increasing the diversity of crop rotation, increasing spring sowing over winter sowing, retaining winter stubble, growing cover crops, managing hedgerows and ponds better, and devoting 10% of the land specifically to wildlife, to provide flower-rich areas in the summer and seed-rich areas for birds in the winter. Organizations such as the Nature Friendly Farming Network in the United Kingdom[‡] and Farming for Nature in Ireland[§] have been set up to promote such practices and to provide a forum in which farmers can learn from each other.

Changes to farming practices in order to encourage wildlife, such as conservation grazing, tend to be directed at maintaining or restoring particular habitats or providing for particular species. In contrast, rewilding aims to restore natural processes that are dynamic and thus lead to continual change. Rewilding is not the abandonment of land: rather it requires active intervention to introduce key species or to remove blockages to natural processes. For example,

---

*Note that animals in conservation grazing systems are generally outside all winter, rather than being housed.

† See https://bit.ly/33LV24Q.

‡ See https://www.nffn.org.uk/.

§ See https://www.farmingfornature.ie/.

one key action of the Wild Ennerdale Project[*] in Ennerdale (a valley in the west of the Lake District in Cumbria) has been to remove a culvert upstream of the lake, Ennerdale Water, replacing it with a single-span bridge. This has restored the natural flow of sediment in the river, creating a dynamic environment in which islands appear and are then washed away. This has improved water quality and increased fish numbers. Three herds of Galloway cattle have been introduced in the valley, and in the places where they have been grazing for more than 10 years there has been a 65% increase in bird species and a doubling of the number of birds.[†]

The Ennerdale Galloway cattle are, in ecological terms, substitutes for the aurochs (wild ox) that once roamed Europe, but they are looked after and, at the end of their lives, they are taken away for slaughter and used for meat, as they would be in a farming system. Similarly, the Knepp Wildland project, covering 3,500 acres in southern England, has introduced a variety of different types of grazing animals, including English longhorn cattle, Exmoor ponies, red deer and a small number of Tamworth pigs, in order to create a dynamic ecosystem of scrub, woodland and grassland on what were, two decades ago, arable fields and dairy pastures (Tree 2018). The cattle and pigs are slaughtered for their meat but, like the cattle of the Morecambe Bay Grazing Company, they are primarily there to perform a role in the ecosystem. The Knepp project now supports a vast array of wildlife, including several species that are in precipitous decline elsewhere in the United Kingdom: in particular the turtle dove, the nightingale and the purple emperor butterfly.

The Knepp Wildland project has been very influential in Britain, not least as a result of Isabella Tree's book *Wilding* (2018). There are many similar large estates that communicate with and learn from each other. The Lowther Estate in Cumbria, for example, has embarked on what looks like a similar strategy, using longhorn cattle and Tamworth pigs, which they are calling Wildland farming. Also in Cumbria, Gowbarrow Hall Farm is developing a hybrid of

---

[*] See http://www.wildennerdale.co.uk/.
[†] See http://www.wildennerdale.co.uk/wildlife/.

regenerative farming and rewilding: in the summer their cattle graze the better land using a mob grazing system and in winter they are allowed to roam freely on a large area of less productive, higher land where there are also some ponies and, in the woodland, a couple of pigs. Gowbarrow Hall Farm aims to restore a woodland pasture ecosystem on this less productive land (Chapman 2020, case studies 4 and 6).

The Knepp project is open to criticism for its impact on those who formerly worked on their farms. In a radio interview, Tree, who owns the estate with her husband, talks about how one of the hardest things they did was to make their farm manager redundant.* One of their sources of income is the letting of residential property previously occupied by their farm workers (Fairlie 2019). In comparison, at the Lowther Estate, the conversion of home farmland from conventional sheep and arable farming to Wildland farming has not been at the cost of jobs, though some roles have changed. Farming for nature does address the biodiversity and climate crises, but in many instances it will provide fewer direct jobs than conventional farming. In marginal areas, though, where farming on the conventional model is going out of business, it can retain some farming jobs, although it is likely to rely on public funding for the environmental benefits it provides in biodiversity, flood prevention and carbon sequestration. It perhaps has the potential to provide many more jobs in nature-based tourism and recreational activities: the Knepp Estate reportedly gets as much income from ecotourism as it does from meat sales or from agricultural subsidies (Fairlie 2019).

A FUTURE FOR AGRICULTURE?

Debates about the future of agriculture are often framed as a question of whether to 'spare' or 'share': to produce the food we need intensively on as small an amount of land as possible, so that other land can be left for nature; or to farm more land in a less intensive

---

* The radio programme Desert Island Discs, interviewing Isabella Tree, writer and conservationist, 29 November 2019 (https://bbc.in/35lvps4).

fashion, thus allowing wildlife to share it with us. This seems to me to be a false choice, primarily because modern intensive farming is not sustainable in its use of energy and other resources and in its impact on air and water quality, so its continuation is not a long-term option. Intensive agriculture depends on fast-growing, highly productive breeds of plants and animals, but the maintenance of these breeds relies on the more genetically diverse traditional breeds to keep the former breeds going. Without them, modern intensive farming would not have the necessary genetic resources when their chemical arsenal eventually fails to stop their animals and plants succumbing to disease.

Instead, farming needs to be done in a way that builds the health of the soil to wean it off dependency on synthetic inputs: these regenerative farming practices are good for wildlife as well as for the profitability of the farm. Regenerative farming has the potential to produce a greater diversity of food from the same farm. There are also opportunities to grow other sorts of crops: fibres such as flax or hemp for textiles and willow or miscanthus grass for energy production. Also, farmers can use their land for renewable energy systems, such as wind turbines and solar photovoltaics, around which at least some farming activities can still take place (for example, in the United Kingdom sheep do well on pastureland with ground-mounted solar farms). On less fertile, poorer-quality land, food production should take second place to the maintenance or restoration of particular habitats, species or natural processes; these objectives may well require particular farming practices, such as conservation grazing, to be carried out.

Regenerative agriculture needs to be supported by policies that build local food economies and that enable farmers to find markets for the diverse mixture of products that a regenerative farming system produces, at the scale it is able to produce them (as opposed to the scale demanded by supermarkets). It would also be aided by a strategy to reduce the chemical arsenal used by agriculture. For example, artificial fertilizers, herbicides and pesticides could be taxed to reflect the damage they cause to soils, the environment and our health. Veterinary medicines such as de-wormers should be less easily available to livestock farms, and their use should require veterinary supervision, so that they are only used when animals are actually sick. These measures would tilt the scales in favour of those farmers who do not use such chemicals.

To support farming for nature we need to find ways to pay farmers for maintaining the wildlife on their land, for holding back flood water, for sequestering carbon and for the other benefits they provide. There are at least two schemes in Ireland that include a 'payment by results' element: the Burren programme and the Hen Harrier Project.* Post Brexit, England is set to transition its farm support payments to the Environmental Land Management Scheme, the principle of which is 'public money for public goods', removing area-based farm support payments.† This has the potential to support farming for nature and nature-friendly farming practices, though much will depend on the details of how the scheme is implemented.

It should be made easier to become a farmer, particularly for those without capital assets, through provision of apprenticeships and making available the tenancies of small farms.‡ There is a need for training in regenerative agriculture, which should include ecology – a subject not generally covered by those studying agriculture. A vibrant, living and working countryside, providing food, space for wildlife and good jobs, will be to the benefit of us all.

Agriculture needs a just transition as much as coal mining communities do, but whereas there is no future for coal mines in a zero-carbon world, there needs to be a future for agriculture.

---

* See https://bit.ly/3sfh4X0 and https://bit.ly/3sgI48Q.

† See https://bit.ly/3HiKeJ4.

‡ The tenancy of a small farm is what enabled Cath and Bill Grayson, mentioned in the 'Farming for nature' section, to become famers. Their farm was owned by the National Trust, a conservation organization that is a large landowner in England. Many County Councils in England used to own a network of small farms that they let out to young and first-time farmers, sometimes at below-market rents, enabling people to get started in farming. Unfortunately, the number of these has halved in 40 years (see https://bit.ly/355F8D5). Organizations such as the Ecological Land Co-operative (https://ecologicalland.coop/) are attempting to fulfil some of this function. They buy up land then split it into smallholding plots for people wanting to set up their own ecological land-based business.

Bibliography

Abram, M. 2020. Regenerative farming: the theory and the farmers doing it. *Farmers Weekly,* 13 October (https://bit.ly/3MiUSDH).

Berners-Lee, M. 2019. *There Is No Planet B: A Handbook for the Make or Break Years.* Cambridge University Press.

Brown, G. 2018. *Dirt to Soil: One Family's Journey into Regenerative Agriculture.* Hartford, VT: Chelsea Green.

Burdett, D. 2020. Regenerative agriculture: making the change happen. Report, June, Nuffield Farming Scholarships Trust (https://bit.ly/3Hv3DXv).

Chapman, A. 2020. A just transition in agriculture. Report, December, Green European Foundation (https://bit.ly/3st5Qym).

Chapman, T. 2012. Are mob grazed cattle the perfect arable break? Report, June, Nuffield Farming Scholarships Trust (https://bit.ly/3C6ksa7).

Clark, C., Scanlon, B., and Hart, H. 2019. Less is more: improving profitability of and the natural environment in hill and other marginal farming systems. Report, November, funded by the RSBP, National Trust and Wildlife Trusts (https://bit.ly/3hrjbRo).

Colantoni, A., Egidi, G., Quaranta, G. D'Alessandro, R., Vinci S., Turco, R., and Salvati L. 2020. Sustainable land management, wildfire risk and the role of grazing in Mediterranean urban–rural interfaces: a regional approach from Greece. *Land* 9(1), 21 (doi: 10.3390/land9010021).

Fairlie, S. 2019. Wild in the Weald. *The Land* 24, 40–42.

Food, Farming and Countryside Commission. 2021. Farming for change: mapping a route to 2030. Report, January (https://bit.ly/3M5e8Eq).

Greenham, T., and Link, M. 2020. Farming smarter: the case for agroecological enterprise. Report, November, Food, Farming and Countryside Commission (https://bit.ly/3K2RHhj).

Hadjigeorgiou, I. 2011. Past, present and future of pastoralism in Greece. *Pastoralism* 1(24) (doi: 10.1186/2041-7136-1-24).

LaCanne, C., and Lundgren, J. 2018. Regenerative agriculture: merging farming and natural resource conservation profitably. *PeerJ* 6, Paper e4428 (doi: 10.7717/peerj.4428).

Prescott, C. E., Grayston, S. J., Helmisaari, H., Kaštovská, E., Körner, C., Lambers, H., Meier, I. C., Millard, P., and Ostonen, I. 2020. Surplus carbon drives allocation and plant–soil interactions. *Trends in Ecology & Evolution* 35(12), 1110–1118 (doi: 10.1016/j.tree.2020.08.007).

Provenza, F., Meuret, M., and Gregorini, P. 2015. Our landscapes, our livestock, ourselves: restoring broken linkages among plants, herbivores, and humans with diets that nourish and satiate. *Appetite* 95(December), 500–519 (doi: 10.1016/j.appet.2015.08.004).

Soil Association. 2018. To plough or not to plough: tillage and soil carbon sequestration. Policy Briefing, November, Soil Association (https://bit.ly/33XZPQM).

Tree, I. 2018. *Wilding: The Return of Nature to a British Farm.* London: Pan Macmillan.

Aviation and the just transition

Robert Magowan

> Workers are prepared to press for the right to work on products which actually help to solve human problems, rather than create them.
>
> The Lucas Plan (1976)

Three arguments have sustained the extraordinary free ride granted to the aviation sector in the global warming era. The first is that its contribution to global warming is relatively small, and will soon be mitigated by technology. The second is that it brings a very modern form of prosperity to swathes of ordinary people. And the third is that its continued growth creates millions of good jobs. This chapter will argue that none of these arguments can be sustained.

If there is to be any progress on the path to a just transition for this sector, three critical correctives – on the limits of technology, on the injustice of hypermobility and on the false hope of employment – must be brought to bear on the industry's playbook of deceit. Policies targeted at rapid and democratic downscaling of the sector currently count the disempowered labour force as an obstacle; however, in future, dedicated transition policies and new bonds of solidarity could facilitate its emergence as an agent of the transition. In the wake of the devastation caused by the Covid-19 pandemic, we can find inspiration in the Lucas Plan: an alternative vision of economic production and workers' roles within it.

IMAGINATION UNDER THE COSH: 'WHAT WOULD YOU DO IF YOU WERE NOT MAKING THAT?'

In January 1975, employees of Lucas Aerospace gathered in a stately-home-turned-trade-union-education-centre near Sheffield,

in England. Their task was self-set: to formulate a plan to combat
the sweeping set of job losses expected to hit the company's highly
skilled, 13,000-strong workforce. At the time, Lucas Aerospace was
one of Europe's largest designers and manufacturers of aircraft systems and equipment. However, like the rest of British industry at the
time, it was facing a new world with new demands.

The defining prefigurations of the neoliberal shift – low growth,
low investment, automation, international competition and a financial crisis – combined to encourage the company, which was largely
dependent on public military contracts, to embark on a programme
of 'rationalization' and to look beyond the United Kingdom for its
labour force. Between 1960 and 1975, the total number of people
employed in the British aerospace industry had already declined from
283,000 to 195,000, and further cuts to government defence spending were privately confirmed. For both the blue- and white-collar
members of the Lucas Aerospace Combine Shop Stewards' Committee (LACSSC), the writing was on the wall.

But the LACSSC decided to go a step further than traditional
union activity. Lucas workers sought inspiration from Italy's Fiat
workforce and the United Automobile Workers of America, who
were expanding the scope of their demands and asserting their
concern about the pollution caused by their industry (from both its
factories and its products). Yet, the aerospace industry faced more
existential threats than this: the technological shift towards more
capital-intensive production was unstoppable. And most unions and
the LACSSC regarded the impending defence cuts as desirable, not
just inevitable.

It was the determination to reckon with these dilemmas head on
that delivered what the Industry Minister of the time, Tony Benn,
called 'the most remarkable exercise in British industrial history'. The
Lucas Plan, as it came to be known, was a 1,200-page document,
drafted and submitted to the company by workers. It detailed more
than 150 ideas for redeploying the skills, labour and machinery used
in the shrinking defence market towards 'socially useful production'.
Wind turbines, lightweight trains and heat pumps were among the
proposals, developed with external academics and innovators and
matched with the existing machinery and workers' skills. Such ideas

were not just technologically ahead of their time, but they also consciously contrasted, in *social* terms, with the current output of the company.

As one of the Lucas Plan's architects, Mike Cooley, wrote: 'We have, for example, control systems that can guide a missile to another continent with extraordinary accuracy, yet blind and disabled people stagger around our cities in much the same way as they did in medieval times.' The Lucas Plan therefore served as both a plan for the workforce's survival *and* a direct moral challenge to the industry's central purpose.

Eulogizing the Lucas Plan can make it sound like a phenomenon firmly of its time: a sepia-toned, almost twee image of workers meeting to deliberate, with ambition and agency, not only the future of their employer but also the nature of technology and progress. If seen this way, however, it is because the space for democratic deliberation and decision making in the modern economy has since become so hollowed out. To even imagine employees behaving in such a way now provokes a feeling of hopelessness and resignation at the sheer naivety of it all.

Other chapters in this book demonstrate the groundswell of policy ambition and practical initiatives to secure jobs and livelihoods in the face of the climate crisis – what we now call a just transition – in sectors like energy, industry and (to a lesser extent) agriculture. But the conspicuous, long-standing absence of aviation is a tale of its own.

AVIATION'S FREE RIDE

Across much of the aviation sector, the very concept of a 'transition' is a misnomer. No mainstream organization of influence, either within the industry or outside it, has yet posited a *reduction* in consumption – a remarkable blind spot at the highest echelons of ostensible climate progressivism. We will consider the sustainability question in the next section, but suffice to say that the UN's flagship Carbon Offsetting and Reduction Scheme for International Aviation entails no aggregate emissions reductions and no demand constraints up to 2050. In more critical quarters, the thinktank Transport &

Environment ('Europe's leading clean transport campaign group') recommends a policy mix that allows for significant demand growth in Europe through to 2030 and 2050.

Europe's expanding just transition policy landscape reflects this peculiar hubris. The EU's flagship €7.5 billion Just Transition Fund, for example, focuses only on the energy transition, with no provisions or funding available to aviation. Similarly, Spain and Ireland's initiatives target only coal and peat production, respectively. Scotland's Just Transition Commission is a much more far-reaching endeavour, covering transport as well as industry, buildings, land and energy. Yet it also fails to mention aviation. This is despite the fact that its final report followed the biggest crisis that the sector had ever faced, with Covid-19 restrictions leading to the loss of 191,000 jobs across Europe. If this is not the time to bring aviation into the just transition picture, when is?

Historically, only fringe advocacy has tempered the absence of aviation just transition policies. The United Kingdom's Public and Commercial Services (PCS) trade union, representing a small proportion of the overall workforce in air traffic control, has been a consistent but lone voice in the sector giving serious consideration to the impact of climate change on the industry and the livelihoods of its workers for close to a decade.

But now, the Covid-19 pandemic has begun, in some corners, to upend the traditional cold shoulder shown towards a serious transition in aviation. For example, although the European Transport Federation's civil aviation section included neither initiatives on a just transition nor any mention of climate change in its 2017–2022 work programme, the trade union made the following statement in September 2020:

> The pandemic might be paradoxically an opportunity to rethink [aviation's] future. As aviation's growth in Europe was to a large extent enabled by social dumping [i.e. shopping around for the lowest employment standards] of low-cost airlines and inefficient state aid for the airports serving such carriers, these practices must be ended.

The union's new head, Eoin Coates, said 'we accept that transport will change, and we are building the principles of a just transition'.

In the United Kingdom, the Trades Union Congress joined climate campaigners to call for a just transition and the 'right to retrain' for workers in the sector, in response to the combined threat of Covid-19, automation and climate change (New Economics Foundation 2020). And in July 2021 Germany's second-largest trade union, Verdi, which represents 2 million workers, took the remarkable step of calling for fewer short-haul flights and a minimum price of €40 per flight (*Der Spiegel* 2021).

In parallel, climate-driven aviation campaigns are intensifying and increasingly incorporating the jobs perspective in their activity. In November 2020 a report entitled 'A green new deal for Gatwick' called for urgent investment in alternative green jobs in response to the decimation of the industry, in this case at the second-largest UK airport (Green New Deal UK 2020). In February, the global Stay Grounded network partnered with PCS to publish a landmark document considering the issues: 'A rapid and just transition of aviation' (Stay Grounded 2021). Inspiration was found in the Broughton manufacturing site of Airbus, which was retooled to help produce ventilators at the height of the pandemic (Wainwright 2020).

All the same, it looks like the opportunity for Covid-19 to catalyse a significant and immediate rethink of the sector will be missed. Though much has been made of the delayed return to pre-Covid demand until 2024, industry voices expect the crisis to dock only 2 years' worth of growth from its long-term expectations (International Air Transport Association 2021).

The absence of a just transition plan for the climate crisis has already failed aviation workers during Covid-19. That failure to prepare has also had disastrous consequences elsewhere: for example, in Scotland, once dubbed the 'Saudi Arabia of renewables', wind turbine construction yards now sit mothballed, and 90% of surveyed oil and gas workers have not even heard of a just transition (Platform 2020). Standing in the way of that transition plan in aviation are three discourses of delay – three playbooks that sustain the aviation sector's exceptionalism. Each will now be discussed in turn.

THE SUSTAINABILITY PLAYBOOK

Just as workers at Lucas Aerospace began to question the 1970s aerospace industry's mission of creating weapons in peacetime, the commercial aviation sector is, in the midst of a climate crisis, increasingly subject to scrutiny today.

Aviation emissions in 2018 amounted to 1 billion tonnes of CO_2: 2.5% of the global total (Ritchie 2020). If the sector were a country, it would be the sixth largest emitter in the world, sitting between Germany and Japan. However, this is before taking into account the 'radiative forcing' effect of non-CO_2 emissions at high altitudes. These effects mean that aviation emissions warm the atmosphere at three times the rate associated with other CO_2 emissions (Lee *et al.* 2021). That effectively triples aviation's impact, raising emissions to the equivalent of more than 3 billion tonnes of CO_2, which is greater than global emissions from cement (2.8 billion tonnes) and closing in on the total amount of emissions produced by the entire EU (3.5 billion tonnes). Military figures are hard to come by, but the 13 million tonnes (of CO_2 equivalent) estimated for the US Air Force indicates that they are not trivial amounts (Belcher *et al.* 2019).

Global demand for flying is expected to double over the next 20 years. The consequences of this growth for global emissions are extreme: even if the industry improves efficiency as much as its best ambitions predict, the sector is projected to emit more than 40 billion tonnes of CO_2, or a fifth of the world's entire carbon budget for 1.5 °C of global warming (Pidcock and Yeo 2016).

This puts aviation on a sharply different trajectory to most other industries. Ryanair's recent entry into the top ten European emitters illustrates the point: aviation might not yet be the new coal, but it soon could be, particularly as emissions from other sectors fall.* The sector's successful self-isolation from the bounds of climate policy, on the basis of its supposedly small contribution, is coming to an end.

---

* See https://bit.ly/35ppdjd.

There is a long history of aviation having a fantastical relationship with ecological reality. In 1973, the Maplin Sands off Britain's Essex coast were proposed for the site of the 'world's first environmental airport'. Yet, plans paradoxically included creating a new eight-lane motorway to London (Beckett 2009, p. 42). In the end, Maplin airport never materialized. Over the course of the aviation industry's growth during the last half-century, similar promises of a sustainable future have been gobbled up by governments and media outlets, hungry for souped-up techno-fixes to the climate crisis. Unfortunately, that vision has turned out to be largely mythical.* The Aviation Environment Federation (2021) puts it most succinctly: 'The aviation industry has yet to bring to market a fuel that releases less CO_2 from the tailpipe of an aircraft than fossil fuel.'

Numerous analyses have found that all legitimately low-carbon aviation technologies are *at least* ten years from delivery at scale; the EU forecasts that large, zero-emission aircraft will not reach the market until 2035. Finlay Asher, a former Rolls Royce aerospace engineer who is now an aviation campaigner, describes the industry's well-rehearsed and trenchant promotion of various techno-solutions as the 'sustainability playbook'. Under pressure to justify its growing emissions, the industry has chosen to 'mislead the public and politicians about the impact of flying'. Let us consider the various technologies in turn.

Perhaps the simplest and most tantalizing technology target is electric flight, given the rapid decarbonization of the energy sector in Europe. In 2019 the launch of a single nine-seater prototype electric plane was all it took for the BBC to declare that 'the age of electric flight is finally upon us' (Bowler 2019). Recent sober assessments, however, find that the technology's potential in the medium term is limited to small planes (under 150 passengers) and, critically, to short-haul flights under 1,500 km, which only account for 20% of aviation emissions (Air Transport Action Group 2021). Even then, the deployment at scale is still a long way off (Beevor and Murray 2018).

---

* See https://bit.ly/3pvq48C.

Another technology with zero-carbon potential is hydrogen. In 2000 Airbus began EU-funded research into hydrogen fuel and even made a commitment to substitute it for kerosene by 2020 (Fitzpatrick 2010). In 2010 the programme was shelved. This has not got in the way of continued public relations success: the BBC published a puff piece entitled 'The hydrogen revolution in the skies' in April 2021, for example (Henderson 2021). However, the technology remains at least a decade away from commercial use, and due to its emission of atmosphere-warming water vapour, the UK Climate Change Committee (2018a) has recommended against it altogether. The EU expects neither hydrogen nor electric batteries to play a significant role in aviation before 2040.

Instead of electric and hydrogen, the industry has channelled sizeable research, development, marketing and lobbying energy into sustainable aviation fuel (SAF), a loosely defined set of technologies based principally on the 'carbon neutrality' of combusting the CO_2 absorbed in biomass growth. 'In the medium term,' according to the International Air Transport Association, 'SAF will be the only energy solution to mitigate the emissions growth of the industry.' However, biofuels (to use the market-unfriendly term) are subject to an emerging global resource race that is already exacerbating environmental destruction and injustice, centring on land grabbing and loss of food sovereignty, as documented by the campaign group Biofuel Watch. As a result, the UK Climate Change Committee (2018b) recommended that biofuels only contribute 10% of aviation fuel by 2050. Synthetic fuels, meanwhile, are at the early stages of development and are significantly more expensive (currently nine times more) than kerosene; they are projected to remain four times more expensive in 2030 (Asher 2021). New EU targets for SAF are just 5% of the fuel mix by 2030 (with 0.7% of that synthetic), rising to two-thirds by 2050 (with 28% of that synthetic) (Carbon Brief 2021).

The upshot of this is twofold. Firstly, the EU's settled choice to combine cheap offsetting with biofuels represents just another form of what Jason Moore calls 'cheap nature': the centuries-old pursuit of accumulation by expropriation of territories and resources (Moore 2016). Secondly, in the meantime, the technological waiting game allows for at least a decade of almost entirely unfettered emissions. Every single flight across the Atlantic in that time will burn through 70,000 kg of (still) tax-exempt kerosene, producing more than three times that weight of CO_2, which will then linger in the atmosphere for up to 35,000 years.

Behind this, and sustaining it, is a series of enormously successful PR campaigns and climate-focused advertising, portraying even the most tentative technological advancements as paradigm shifts and adding increasingly tenuous buttresses to the industry's social licence to operate. Partnerships with fossil fuel companies, such as the deal struck by Rolls Royce with Shell, help prop up the social license of that industry as well (Reuters 2021). A study by InfluenceMap (2021) found that intense lobbying efforts sell a vision of a highly innovative industry with a prosperous future to policymakers, while at the same time opposing demand management policies like kerosene fuel taxes, the inclusion of aviation in the EU Emissions Trading System and ticket taxes on short-haul flights.

Despite a decades-long exercise in corporate 'greenwashing', the vision of a low-carbon aviation sector is a long way from materializing. This is the first major injustice of the aviation industry: the havoc it is wreaking with planetary health and human societies, for generations to come.

THE PROSPERITY PLAYBOOK

This deep injustice of aviation's polluting impact is only matched by its social injustice. For all the industry's talk of delivering cheap thrills to the masses, its benefits are incredibly unequally spread. More than 80% of the people on earth are said to have never set foot on a plane (Gurdus 2017). A mere 1% of the global population is responsible for more than 50% of aviation emissions. A jaunt from London to Rome adds more CO_2 to the atmosphere than the total amount emitted by an average citizen of Nepal, Haiti or some twenty-eight other countries in an entire year (Kommenda 2019). Even in the United Kingdom, and contrary to the popular perception of the summer holiday abroad being the model of a rest-and-recuperate leisure activity, more than half the population do not fly in a given year. A staggering 70% of flights are taken by just 15% of the population (those who fly more than three times a year).

Legal rights as well as financial access are also denied to many, creating a 'global mobility divide' (Mau *et al.* 2015). While the average European can travel to sixty-two countries without a visa, the average African citizen can travel to just fifteen. Vapid arguments for the industry's expansion in the Global South ignore the 300-plus cases of socio-environmental conflict generated by airport projects and relating to 'land acquisition, displacement of people, destruction of ecosystems, local pollution and health issues', as documented by the Environmental Justice Atlas (2020).

The concentrated consumption in the aviation sector, and its disproportionate claim to scarce resources, adds to the broader criticism levelled by Ulrich Brand and Markus Wissen against the 'imperial mode of living' (2021). This concept describes the incorporation of everyday, unconsciously reproduced activities in a capitalist production model that appropriates resources, pollutes sinks and exploites labour. Despite flying being a service used mainly by very concentrated social groups, its social licence, and crucially its ongoing expansion, is based on the *presentation* of flying as an everyday behaviour, its thorough embedding in cultural and social norms, and its insertion into widespread but simple forms of aspiration.

The Covid-19 pandemic clearly highlighted the connections between the inequity of aviation and its ecological impact. As Andreas Malm (2020) wrote, the virus itself originated in 'zoonotic spillover' driven by the expansion of capital into wild habitats. Its transmission then followed aviation lines, 'giving rise to the paradox that rich people were the first to contract the virus'. Elite tourism locations such as Austrian skiing resorts – normally getaways for wealthy Europeans – became epicentres of transmission and superspreaders to home locations across the continent. While the vast majority of commercial flights remained grounded throughout 2020, the number of private jet flights returned to pre-Covid levels in the summer of that year. There is a troubling parallel here with the function of aviation in global warming: though private jets and frequent flyers do not transport the climate crisis from one place to another, in the absence of punitive restrictions, every journey they take extends and intensifies it.

One lesson we might learn from the pandemic, in order to challenge the social licence of the aviation industry, is to push for a reframing of its problematique away from 'sustainability' – something that is increasingly promoted by industry players. As it stands, the territory on which stakeholders increasingly compete is around this trajectory of increasing sustainability. But as Sharachchandra Lele (2020) has said, discourses of sustainability helpfully obscure injustices in the here and now. If someone in the Global South is asked why it is wrong that they experience flooding exacerbated by frequent-flyer emissions or higher food prices caused by aviation's biofuel race, they are likely to say 'Because it's unfair', not 'Because it's unsustainable'.

We do not have to follow much more of the thread (from production to consumption to impacts), as the workers of Lucas Aerospace did for the defence industry in the 1970s, to see that aviation workers today find themselves in the service not of distributed prosperity, but of unequal access, concentrated harm and spatial injustice. Startlingly few people are benefitting from the production of a shameful degree of pollution – a gross misallocation of precious atmospheric space. Though few are prepared to say it yet, the industry's urgent degrowth is a critical precondition of any just transition.

THE JOBS PLAYBOOK

If the prospect of decline aligns today's aviation workforce with defence aerospace manufacturers in the 1970s, it is its relative weakness in the economy that sets it most obviously apart. Trade union density has been in general decline since 1980: from more than 50% of the workforce in many countries to just over 20% on average across the EU now (Vandaele 2020). Into this power vacuum stepped a liberalized, newly developing industry in a globalized, expanding economy: a remarkable and deep-seated process that has accompanied the sector's explosion over the last 30 years. The 'jobs playbook' that the industry has been able to deploy as a result is a combination of mistruth, mistreatment and misappropriation, but above all it reflects political and cultural power. It allows the industry to seek expansion in the name of workers for whom it shows little regard.

Overpromise and exaggerate

The first move in the jobs playbook is to overpromise and under-deliver. Despite the industry having trebled in size since the early 1990s, employment levels in Europe have remained stable and have declined in some sub-sectors (European Parliament 2016). Such realities have not prevented the sector from being represented as a job-creating machine. In 'Airport jobs: false hopes, cruel hoax', a 2013 report for the Aviation Environment Federation, economist Brendan Sewill explored the means by which the industry exaggerated its employment contribution in the present, in order to support its case for expansion. Much more recent work from the New Economics Foundation has further compounded the unreliability of industry employment projections.

On aggregate, the industry in Europe claimed that before Covid-19 the air transport sector supported 13.5 million jobs. But underneath such headline figures lie all manner of ills. Only 2.7 million of those are in 'direct' aviation employment, and of these, only 1.3 million are in what we would understand to be aviation employment. The other 1.4 million work at airports but in retail outlets, restaurants and hotels. Such figures obscure the simple displacement of these jobs from the high street to the airport – a shift naturally exacerbated by duty-free retail.

Of the remainder claimed by the industry, 3 million come from supply chain jobs and 2.2 million from wage payments to directly employed staff – both knock-on effects that would occur (albeit to varying degrees) regardless of the initial employment type. But it is the final contingent of 5.6 million that is the figure that is most routinely misused and that reveals a critical weakness in the case for an airport's contribution not only to jobs but also to economic prosperity. This figure represents the impact of tourism on European soil, with arrivals purchasing goods and services and stimulating employment elsewhere in the economy. But this accounts for only one side of the equation: tourists *from* the home region or nation also take their spending power abroad. To overlook this 'two-way road' fatally undermines the case for airport expansion, nationally and locally. A critical analysis of the application to expand Leeds Bradford Airport, for example, found that its total economic contribution would actually be negative once such factors were taken into account.

Finally, business travel, contrary to popular perception, is a small and shrinking part of the modern commercial aviation market. In the United Kingdom it peaked in 2006, and of course it has been upended by Covid-19 as companies find new (quicker and cheaper) ways to communicate. Airport forecasts – and the image of airports as harbingers of foreign investment and plucky entrepreneurial interaction – are yet to adapt.

Exploit weak workforces

Underlying the statistical manoeuverings are the structural characteristics of automation and efficiencies of scale – factors that have both kept job growth to a minimum and made workers notoriously vulnerable to cycles of boom and bust. While employment in the sector tends to grow slowly during periods of expansion, it shrinks rapidly during the sharp and frequent contractions in demand that result from events such as the Gulf War and the Global Financial Crisis. Research from the New Economics Foundation showed that following the latter (the most recent crisis before Covid-19), passenger levels recovered by 2013 and had grown by 30% by 2019 in the United Kingdom, but employment levels never actually got back to their pre-crisis peak.

Covid-19 has revealed this tactic at its most brazen. Despite the industry receiving a total of €38 billion in bailouts from the EU (Transport and Environment 2021), and £9 billion from UK public support, 191,000 and 60,000 aviation jobs, respectively, have been shed. Airports and airlines alike lost no time in announcing redundancies, many before lockdowns had even begun and before employment stabilization programmes like the United Kingdom's furlough scheme had even been introduced. Lufthansa, for example, received €9 billion last year from the German government's emergency Covid-19 bailout but laid off 60,000 employees worldwide. In the United Kingdom, meanwhile, the 'fire and rehire' tactic, which is illegal in other European countries, was used opportunistically by leading airlines such as British Airways to weaken the pay and conditions of staff.

Rather than keeping on or rapidly rehiring former employees in recognition of public intervention, the sector has demanded further financial support and policy provisions to ensure flights get back off the ground as quickly as possible, putting aviation into conflict with public health measures that are designed to prevent the spread of Covid-19. It has done so in the name of jobs, arguing throughout tight lockdowns across Europe in winter and spring that coordinated provisions for international travel were essential 'to save jobs and [the] upcoming summer season'. Desperate UK trade unions such as the GMB rowed in behind, aligning with airlines' shameless calls for Air Passenger Duty to be frozen. In a telling indicator of potential cleavages within unions, however, the campaign work of Unite, the United Kingdom's largest union, has a slightly different slant. Despite facing the loss of over 5,000 jobs a month, Unite rejected the pausing of Air Passenger Duty as a short-term solution.

DEMOCRATIC DEGROWTH

What constitutes a just level of aviation, then, and how is to be reached? What approach meets society's needs *and its desire* for mobility without sacrificing large numbers of livelihoods for the hypermobility of a few?

While the calls to moderate flying – to constrain its growth and to better distribute it – present an important challenge to aviation's central injustice, they are not sufficient. If everyone on earth took just one short-haul return flight per year, global aviation emissions would approximately double (Rutherford *et al.* 2019). One long-haul return flight a year would increase emissions tenfold. There is therefore, unfortunately, no model of mass, regular aviation in the near future that can avoid the destruction caused by such emissions levels.

Before Covid-19 there were already signs of what we might call a consumer-focused 'perceptive adjustment' in regards to the aviation sector – a tentative attack on its greenwashing and its privileged status. The late 2010s saw louder calls to voluntarily fly less and to reject the industry's desperate promotion of jet-setting business and holiday lifestyles. In Sweden this cultural shift inspired the name *flygskam* – 'flight shame'; flight passenger numbers in Sweden were already falling before Covid-19, while rail numbers were increasing. The rapid social and economic transitions of the past show that these behaviour changes can stimulate and snowball further action via a combination of voluntary shifts in practices and norms, private sector responses to demand for alternatives, and facilitatory public policy (Simms 2018).

Europe's night train revival is a good example. Of the 365 cross-border rail links that once existed in Europe, 149 were no longer operational in 2018 and only 57 were fully exploited (Donat *et al.* 2020). Short-haul flights gradually replaced the luxury heyday of the Trans-Europe Express in the 1960s, reducing rail to just 8% of all passenger travel in EU member states. But domestic initiatives spearheaded by publicly owned rail operators in Austria, Germany, France and Switzerland have led to dozens of new routes in recent years. An aligned policy shift is (slowly) emerging, with several European governments (and the EU) moving to reduce or end aviation industry tax exemptions (Valero and Baiter 2019). In April 2021 the French government took the unprecedented decision to ban short-haul flights where the same journey could be made by rail in under 2.5 hours, watering down the 4 hour recommendation of the Citizens' Convention on Climate (BBC 2021).

However, it is folly to be complacent when viewing this transition's inevitability. Modal shift only offers an attractive lever for the one-fifth of emissions that come from short-haul flights. It offers no guarantee of the broader transformation required – what we might call a 'mode of living shift', from 'boundless to conscious mobility' (Stay Grounded 2018). No expectations should be placed on ecological determinism: the assumption that environmental limits will operate as boundaries against which flying behaviours and policies will suddenly recoil, 'by design or by disaster'. Yes, the sector's ambitions to double passenger numbers globally by 2037, and to soar harmlessly using the fuels of the future until 2050, can read preposterously: for the climate-versed, those years are mere etchings of ascending degrees of climate disaster. But the accumulative and expansive drive of capitalism makes it incredibly resilient – able to avail itself of both rapid technological advancement and enormous political and cultural power to stave off significant threats that may topple incumbents. This is the story of the fossil fuel industry, expanding in spite of 'peak oil' and 'stranded assets' for decades, and still the recipient of $2.7 trillion from lenders globally since the Paris Agreement was signed (Rainforest Action Network 2020).

Although the Global Financial Crisis saw predictions abound that the aviation sector was in permanent decline, the 2010s ended up being its most successful decade in history. Even post-Covid-19, investment in expansion proliferates, with more than £20 billion currently planned in UK airports alone, and with similar figures for India and the United States (Green House Think Tank 2021), and an estimated $2.4 trillion demanded globally by 2040.

In reality, high-carbon aviation, especially in the absence of near-term technology revolutions to displace it, can only really decline in line with the balance of social and political forces that are brought to bear on it. As it stands, those forces consist of the major airlines, aerospace manufacturers, trade associations cum supranational government agencies representing the industry's interests (the International Air Transport Association and the International Civil Aviation Organization), and national governments and lobby groups. The industry's mistreatment of workers throughout Covid-19 – such as proliferating redundancies and weakening their pay and conditions – testifies to the fact that trade unions play a minor role in that constellation.

All the same, a transition to 'climate just mobility' can still be envisaged. In practice, it can only be the consequence of a ruthless policy mix, applied not in deference to the pace of technology development but in line with the urgency of degrowth. Measures could include a moratorium on all airport expansion in Europe and strict limits elsewhere,* significantly higher taxes on kerosene to raise the relative cost of flying (much higher than current recommendations),

---

* See https://bit.ly/3tlc6r7.

a frequent-flyer levy to target the highest emitters and to facilitate travel by the lowest emitters (New Economics Foundation 2021), an outright ban on fossil-fuelled private jets (A Free Ride 2019), and a sharp reduction in military budgets and aviation's role therein.

The objective of such a policy mix might in practice look like what we see in figure 1: a rapid decline in global passenger kilometres to a fraction of their current level over the next decade. This would be driven above all by policies targeting the cessation of hypermobility. But at the same time, a degree of 'cultural overspill' could simultaneously cause *infrequent* flyers to reduce consumption in the next decade and to seek alternative modes, speeds and destinations for travel and leisure. Meanwhile, a reassessment of 'imperial mode' tourism and a corresponding rebalancing in local environmental-justice conflicts could also constrain expansion in the Global South. A modest, *widely distributed* increase in flights could then occur from 2030 onwards as zero-carbon technologies become increasingly available.

As Kenta Tsuda wrote, 'It is one thing to choose to live by limitarian ethics, another to legislate it' (2021). What constitutes 'just' aviation is of course complex and contingent on both technology and democratic deliberation. We might now sketch some possible openings for a frontline role for workers in these deliberations, breaking down the barriers to solidarity between anti-aviation campaigners and aviation workers, with a view to achieving a rapid and fair downscaling.

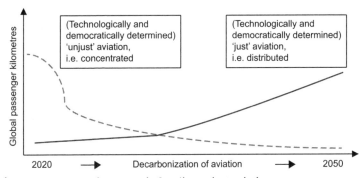

Figure 1. A consumption scenario for climate just aviation.

POLITICAL OPENINGS TO A JUST TRANSITION

Job creation

Green New Deal-type campaigns and policy development have platformed the extraordinary job creation potential of rapid economy-wide decarbonization. As a local example, a report on Gatwick Airport showed that an urgent programme of investment across public transport, building retrofits and care work would more than make up for the huge job losses sustained at the airport as a result of Covid-19 (Green New Deal UK 2020). The charity Possible has gone further and modelled the potential jobs impact of rapidly downscaling UK passenger numbers. It found that if flights were reduced by two-thirds, almost 200,000 UK aviation jobs would be lost (Meadway 2022). However, if replaced by a mix of international rail, ferries and domestic tourism – so no reduction in the overall number of journeys – there would be a net gain of over 300,000 UK jobs.

This hints at the potential for the whole transport sector to leap from decarbonization laggard to leader – *driving* the re-localization of economic activity instead of facilitating its export. The groundswell of domestic tourism in 2020 and 2021, initially enforced by the limitations of Covid-19, could be encouraged and capitalized on by using politically potent advocacy of domestic leisure as a font of community wealth and local job creation. Trade unions that reach across transport sectors are well placed to support the transition of workers from one mode to another while maintaining salaries and job fulfilment. Furthermore, subsidies for airports and aeroplanes as 'skyhooks of capital' could be redistributed towards local transport and leisure infrastructure, creating sites of public luxury that could even include redundant airport space itself.

A broader redeployment of public investment, not just in the transport sector but also in socially critical, labour-intensive (and low-carbon) sectors like care, would help reorient national and local economies and would contribute to a vital improvement in the quality of work in these career paths.

Right to retrain

The solid protection of people's livelihoods is a crucial component of a just transition. In the case of aviation, a right to retrain would accommodate the debts incurred by workers such as pilots; guarantee a commensurate minimum salary both for retraining and for future employment; and offer a range of options accounting for the diversity of both ambition and employment opportunities in economic decarbonization. The recent Possible report found that over two thirds of UK aviation workers would consider seeking employment outside the sector, identifying access to retraining and financial support while retraining as the most important enabling factors (Meadway, 2022). Isolated initiatives under Covid-19 hint at the potential: for example, pilots retraining to become train drivers and flight attendants moving into rail customer service and care work. A group of aerospace engineers in France (Supaero Decarbo) proposed an 'industrial alliance for the climate' in a recent report, in order to oversee the transition (2021). Concluding that a short-term decline in jobs was the most likely scenario if the sector is to stay within carbon budgets, they proposed that such a body would take charge of 'reallocating the production capacities currently underutilized to produce the equipment needed for the energy transition'.

However, as yet, insufficient thought has been put into redeploying the vast array of skills required to run an airport, all with considerable potential. The role of trade unions and worker representative organizations will be an essential corrective.

With workers, against airlines

Common ground exists on key issues between aviation trade unions, workers and environmental campaigners, and cleavages with airlines can be exploited to highlight the endemic injustice of the sector even to the detriment of those platformed as its biggest beneficiaries. 'Social dumping' is a major concern, for example, with growing numbers of pilots flying without direct employment, especially at low-cost airlines. The argument can and should be made that this is bad for both people and the planet: the lack of current protection only exacerbates the future threat that a just transition could carry for job security. Policies like 'fire and rehire' do the same, driving down conditions and morale, weakening the bargaining power of employees and extinguishing the political and mental space for long-term ambition. Cost-cutting as a general policy strategy (for both governments and companies) is a similarly damaging characteristic of the sector, and looks set to get worse through the EU's Single European Sky policy.

Ending unstable employment is already a priority objective of trade unions, and there is every reason for climate activists to offer full support. A significant redirection in aviation economic strategy, with public ownership at its heart, would allow disparate objectives of worker protection, technology development, demand constraint and modal shift to be delivered strategically, 'under one roof'.

CONCLUSION

The aviation industry relies on its playbooks of sustainability, prosperity and jobs for its licence to pollute for the benefit, overwhelmingly, of a wealthy few. As the employees of Lucas Aerospace showed, however, the most powerful voices against this kind of injustice will be those whose labour engineers and services it every day. Aviation workers today, despite being disempowered, unprotected and thrust into crisis after crisis, are still well placed to help break the social licence of their employers and to act instead as critical agents of a just transition for the sector.

Bibliography

Air Transport Action Group. 2022. Fact and figures. Website (https://bit.ly/3q5X3kp).

Asher, F. 2021. Alternative jet fuels. YouTube, Green Sky Thinking, 13 January (https://www.youtube.com/watch?v=XNgmKyw4qfo).

Aviation Environment Federation. 2021. Industry's proposed interim aviation climate targets: AEF comments. Press Release, 22 June (https://bit.ly/3wKYB7v).

BBC News. 2021. France moves to ban short-haul domestic flights. *BBC News*, 12 April (https://bbc.in/3q23lBB).

Beckett, A. 2009. *When the Lights Went Out: Britain in the Seventies.* London: Faber & Faber.

Beever, J., and Murray, L. 2018. Electric dreams: the carbon mitigation potential of electric aviation in the UK air travel market. Report, November, Fellow Travellers (https://bit.ly/3u0ZOV8).

Beever, J., and Murray, L. 2019. Jet, set, go: the case for electric-only UK private jet flights from 2025. Report, November, Fellow Travellers.

Belcher, O., Bigger, P., Neimark, B., and Kennelly, C. 2019. Hidden carbon costs of the 'everywhere war': logistics, geopolitical ecology, and the carbon boot-print of the US military. *Transactions of the Institute of British Geographers* 45(1), 65–80 (doi: 10.1111/tran.12319).

Bowler, T. 2019. Why the age of electric flight is finally upon us. *BBC News*, 3 July (https://www.bbc.co.uk/news/business-48630656).

Brand, U., and Wissen, M. 2021. *The Imperial Mode of Living: Everyday Life and the Ecological Crisis of Capitalism.* London: Verso Books.

Carbon Brief. 2021. Q&A: how 'Fit for 55' reforms will help EU meet its climate goals. Website, 20 July (https://bit.ly/3MpmI15).

Climate Change Committee. 2018a. Hydrogen in a low-carbon economy. Report, November (https://bit.ly/3vCxyuw).

Climate Change Committee. 2018b. Biomass in a low-carbon economy. Report (https://bit.ly/38bpqYj).

Der Spiegel. 2021. Ver.di fordert Mindestpreis für Flugtickets. *Der Spiegel*, 7 July (https://bit.ly/3pw7b5F) (In German).

Donat, L., Treber, M., Janeczko L., Majewski, J., Lespierre, T., Fosse, J., Vidal, M., and Gilliam, L. 2021. Hop on the train: a rail renaissance for Europe. Briefing, 20 January (https://bit.ly/3psJivW).

Environmental Justice Atlas. n.d. Map of airport-related injustice and resistance. Website (https://bit.ly/3IJZOPJ).

European Parliament. 2016. Employment and working conditions in EU civil aviation. Briefing (https://bit.ly/3JNOiDb).

Fitzpatrick, M. 2010. Aviation industry 'ditches' hydrogen. *BBC News*, 17 November (https://bbc.in/36D1vAr).

Graver, B., Zhang, K., and Rutherford, D. 2019. CO2 emissions from commercial aviation. Website, 19 September, International Council on Clean Transportation (https://bit.ly/3J1reQx).

Green House Think Tank. 2021. Global public investment requirements for zero carbon. Report, 1 October (https://bit.ly/3iNAxZz).

Green New Deal UK. 2020. A green new deal for Gatwick. Press Release, 25 November (https://bit.ly/3hzcBbM).

Gurdus, L. 2017. Boeing CEO: over 80% of the world has never taken a flight. Website, 7 December, *CNBC* (https://cnb.cx/3IPI3Oa).

Henderson, C. 2021. The hydrogen revolution in the skies. *BBC News*, 8 April (https://bbc.in/35nj9ru).

Influence Map. 2021. The aviation industry and European climate policy. Report, 1 June (https://bit.ly/3Ltm2qd).

International Air Transport Association. 2021. Brian Pearce – IATA 20 year forecast. YouTube, 17 May, IATA (www.youtube.com/watch?v=5tI_WNySQX0&t=7s).

Kilipiris, E. 2021. Demand for air transport employees surges as industry recovers from pandemic. *Airways Magazine*, 28 March (https://bit.ly/3iCZoyZ).

Kommenda, N. 2019. How your flight emits as much CO2 as many people do in a year. *Guardian*, 19 July (https://bit.ly/3Lk9wsU).

Lee, D. S. *et al.*. 2021. The contribution of global aviation to anthropogenic climate forcing for 2000 to 2018. *Atmospheric Environment*, 244, Paper 117834 (doi: 10.1016/j.atmosenv.2020.117834).

Lele, S. 2020. Environment and well-being. *New Left Review* 123 (May/June) (https://bit.ly/3qJaPtk).

Malm, A. 2020. *Corona, Climate, Chronic Emergency: War Communism in the Twenty-First Century.* London: Verso.

Mau, S., Gülzau, F., Laube, L., and Zaun, N. 2015. The global mobility divide: how visa policies have evolved over time. *Journal of Ethnic and Migration Studies* 41(8), 1192–1213 (doi: 10.1080/1369183X.2015.1005007).

Meadway, J. 2022. The right track for green jobs: cutting aviation emissions while boosting employment and climate-friendly travel. Report for Possible, 20 February (https://bit.ly/3tMi4CO).

Moore, J. W. 2016. The rise of cheap nature. In *Anthropocene or Capitalocene*, pp. 78–115. Oakland, CA: PM Press.

New Economics Foundation. 2020. At least 70,000 jobs in aviation and aviation supply chains at risk. Press Release, 10 June (https://bit.ly/36WbqB7).

New Economics Foundation. 2021. A frequent flyer levy. Report (https://bit.ly/3wM5llv).

Pidcock, R., and Yeo, S. 2016. Analysis: aviation could consume a quarter of 1.5 °C carbon budget by 2050. Website, 8 August (https://bit.ly/3uySn7X).

Platform. 2020. OFFSHORE: oil and gas workers' views on industry conditions and the energy transition. Website, 29 September (https://bit.ly/3wa89bU).

Rainforest Action Network. 2020. Banking on climate change 2020. Press Release (https://bit.ly/3IPoZj9).

Reuters. 2021. Rolls-Royce partners with Shell in sustainable aviation fuel push. Website, 30 June (https://reut.rs/3pylXJg).

Ritchie, H. 2020. Climate change and flying: what share of global CO2 emissions come from aviation? Website, 22 October (https://bit.ly/3uAddnu).

Sewill, B. 2009. Airport jobs: false hopes, cruel hoax. Report, Aviation Environment Federation (https://bit.ly/34c4dvC).

Simms, A. 2018. Climate and rapid behaviour change. What do we know so far? Report, 15 October, Rapid Transition Alliance (https://bit.ly/3Dj0n11).

Stay Grounded. 2021. A rapid and just transition of aviation. Report, 3 February (https://bit.ly/347bIEd).

Supaero Decarbo. 2020. Flying in 2050: what aviation in a constrained world? Report (https://bit.ly/3vyCCjB).

Transport and Environment. 2021. Bailout tracker. Website (https://bit.ly/3NuuyHb).

Tsuda, K. 2021. Naive questions on degrowth. *New Left Review* 128 (March/April) (https://bit.ly/3uzcK53).

Valero, J. and Baiter, K. 2019. Emission-busting aviation taxes gain ground, despite airline pushback. Website, 15 July, Euractiv (https://bit.ly/3vNPrqz).

Vandaele, K. 2020. Bleak prospects: mapping trade union membership in Europe since 2000. Website, 5 November, European Trade Union Institute (https://bit.ly/3MfpFB6).

Wainwright, H. 2020. From airplane wings to ventilator parts. Website, 27 April, New Lucas Plan (https://bit.ly/3w7aFzC).

PART V

Concluding Remarks

To conclude: our work has only just begun

Dirk Holemans and Adrián Tóth

In this book we have tried to show the history, relevance and urgency of the concept of just transition. It is clear that just transition cannot be reduced to a single policy mechanism, legislative initiative, linear process or financial instrument: it has to serve as a guiding compass, feeding into all policy domains, whether we are talking about agriculture, energy, the economy, health, trade or education.

One of the lessons we can learn from the book is that just transition implies quite different challenges in the different regions of Europe. The capacities and resources of, for example, a high-income Scandinavian welfare state that is characterized by low inequality are incomparable with those of some low-income Eastern European countries with limited welfare institutions and a high level of inequality. This shows that solidarity mechanisms organized at EU level are key, without denying that the challenges are still immense even for high-income countries.

The good news is that awareness of the need for solidarity is growing. In the light of the ambitious Green Deal, the European Commission launched its Sustainable Europe Investment Plan in January 2020, including the newly set up Just Transition Mechanism, which was to make at least €100 billion available to support workers and citizens in the regions most impacted by the transition. The first pillar of this mechanism, the Just Transition Fund, has been quintupled to €40 billion as part of the Next Generation EU recovery plan, which is the European Commission's answer to the detrimental impacts of the Covid-19 crisis.

One year later, in July 2021, Europe's ambition to become the world's first climate-neutral continent by 2050 was given extra support through a legislative proposal from the European Commission known as the 'Fit for 55' package. This includes a proposal to implement a system of carbon pricing (the emissions trading system) for the heating and mobility sectors. Yet, it is clear that this could harm low-income groups that already find it difficult to make ends meet each month. Due to rising criticism, the Commission proposed a new Social Climate Fund that mobilized around €72.2 billion for a fair transition. This sounds like a big amount, but it is surely not enough to build a social protection floor underneath the necessary climate policies. Nevertheless, we should not be cynical or pessimistic: it is part of the ongoing societal and political struggles and fights to make sure that the fund, along with other mechanisms, becomes strong enough to make just transition a reality.

It would be unwise to limit our concluding remarks to Europe or the EU. We deliberately did not restrict the contributions in this book to the different European perspectives but instead gave a platform to views from other continents as well. This is partly because climate change, just like the Covid-19 crisis, does not stop at borders (although the impacts can be felt in very different ways depending on geography), but it is also the case that there are other compelling reasons for avoiding the risk of a Eurocentric view.

The first is the historical responsibility of Europe. It is clear that in recent years, China and the United States have been the biggest polluters in the world. But if we look at Europe's historical environmental debt, we quickly realize that we have also employed unsustainable and inequitable colonial practices that have been going on for centuries. Europe is responsible for stripping communities all over the world of their natural heritage and their wealth of resources. This, in turn, shaped Europe. As Jason Hickel (2017) has said: 'Europe didn't develop the colonies. The colonies developed Europe.'

And this brings us to the second reason: the responsibility that we currently have because we have not yet changed the structure of our extractive economies. Moreover, we have to look carefully at our plan to move towards 100% renewable energy (one of the EU's priorities) as part of the just energy transition via the European Green Deal. This plan is of course good and necessary, but wind turbines, solar panels and batteries for energy storage all require large amounts of metals. Where will these metals come from? How fast will we move towards a circular economy? Who will bear the environmental costs of these extractive economies? These are only some of the questions we need to ask ourselves while also reflecting on our past mistakes.

In political ecology – which reconnects nature and the economy and which poses crucial questions about power relations – historical and actual *ecological debt* are crucial concepts, because they recognize that colonization has not only resulted in a loss of culture, language and a way of life for indigenous peoples, but it has also shaped the world economy into one that monetizes and commodifies the environment (Polanyi 1944). As several planetary boundaries have already been transgressed, the only way to provide the necessary space for the inhabitants of low-income countries to fulfil their needs is to share our planetary resources more fairly. As scholars like Jason Hickel and Julia Steinberger have made clear, it is perfectly possible to allow all the people on this planet to thrive, if we embrace the concept of self-sufficiency. This means focusing on what is enough to live in dignity, instead of endless consumer preferences – or, as Ghandi once said, 'The world has enough for everyone's need, but not enough for everyone's greed.'

A just transition in Europe therefore has to be developed as a shared global responsibility. It is clear that long-term cooperative action is critical if there are to be meaningful results and structural change. Roman Krznaric, the author of the book *The Good Ancestor*, argues that even in this current period of the Covid-19 crisis, we need long-term plans to deal with the challenges we are facing – challenges such as climate change and biodiversity loss. In the next 200 years alone, tens of billions of people will be born, among them our grandchildren, their descendants and their friends and communities. The real question remains: how will these future generations look back on us and the legacy that we are leaving for them?[*] According to the immunologist Jonas Salk, we need to think about the consequences of our actions beyond our lifetimes: 'Rather than thinking on a scale of seconds, days, and months, we should extend our time horizons to encompass decades, centuries, and millennia'

---

[*] See Roman Krznaric's 2020 TED Talk entitled 'How to be a good ancestor', 23 October (www.youtube.com/watch?v=61hRq0D8Zcs).

(cited in Krznaric 2020). We have to integrate this dimension into just transition plans and policies: an exercise in learning to think seven generations ahead.

In this light, young people are certainly at the forefront of long-term planning, because they feel that their future is directly impacted by the climate crisis. They are worried not because of climate alarmism but because the effects of the crisis are already here and are damaging and hurting communities globally. The topic has been targeted by young people for many years, yet the latest generation of activists are louder and more globally connected than their predecessors. Furthermore, they are taking action against the different powers in our democracies: they are putting pressure on governments and parliaments using a wide array of actions in public spaces (such as, of course, school strikes), while also taking advantage of juridical possibilities. For instance, Germany's Federal Constitutional Court ordered the government to expand a 2019 law that aimed to bring the country's carbon emissions down to close to zero by 2050, concluding that the legislation was not enough to ensure the safety of future generations. This was a watershed moment in the fight against climate change as it underscores intergenerational justice and could become a benchmark for future lawsuits (Eddy 2021). Crucially, the court judged that the government has a *duty of care* towards the younger generations: an argument that was also present in other countries such as the Netherlands and Belgium (e.g. Climate Case). These cases open up new possibilities to enrich the conversation about what a just transition means, by interweaving it with the ethics of care (Holemans *et al.* 2022).

This brings us to another crucial point. Young climate activists rightly insist on participating. It would be paradoxical to decide on policies for the future without having the generation of the future around the table. More generally, we can also assert that top-down just transition policies will not work. Politicians have to take responsibility and make bold decisions, but within a framework of real participatory processes and institutions. This is of course a major challenge for the EU as a supranational institution. Direct cooperation with cities, municipalities and local communities could be key here, but regular cooperation with member states is also vital. One crucial challenge is how to make sure the huge amounts of money made available in the different EU funds really go to the most vulnerable households and groups, in order to support them through this necessary just transition of our society.

It is clear that just transition is an important guiding principle to realize the goal of a Europe that is more democratic, more resilient, more socially just and also greener. As we have made clear in this book, we stand before a socioecological transformation that will have an impact comparable to that of the industrial revolution – a period when we reorganized our society on the basis of fossil fuels and the global extraction of resources. Now, we are on our way to a socially just circular economy and a society that runs on renewable energy. To make this transformation a success, the EU has to develop appropriate instruments and make available sufficient funds to make it happen in an equitable way, ensuring that the fundamental rights of all are guaranteed in an inclusive society. This much-needed alternative economic model will necessitate many far-reaching changes, such as the creation of a fair and just taxation system. We are faced with a difficult task: to move forward quickly but without leaving anyone behind.

Bibliography

Eddy, M. 2021. German High Court hands youth a victory in climate change fight. *New York Times*, 30 April (https://nyti.ms/3KjZ6cn).

Hickel, J. 2017. Enough of aid – let's talk reparations. *The Guardian*, 6 October (https://bit.ly/3vHQqbw).

Holemans, D., Osman, P., and Franssen, M. 2022. *Dare to Care. Ecofeminism as Source of Inspiration*. Brussels: Green European Foundation (forthcoming).

Krznaric, R. 2020. *The Good Ancestor: A Radical Prescription for Long-Term Thinking*. New York: Experiment.

Polanyi, K. 1944. Habitation versus improvement. In *The Great Transformation: The Political and Economic Origins of Our Time*, pp. 35–44. Boston, MA: Beacon Press.

About the editor and authors

ABOUT THE EDITOR

Dirk Holemans is the co-president of the Green European Foundation (GEF) and a co-founder and director of Oikos, the green Flemish think tank that aims for social-ecological change. Building a collaborative ecological economy is at the heart of the organization's activities. It researches and inspires new citizens' initiatives. He is also editor-in-chief of Oikos's eponymous journal. He has previously been a member of parliament (in Flanders) and a city councilor (in Ghent) for the green party. He has worked as a researcher and lecturer at different universities in Europe. His main interests are ecological economy, commons, green cities and political ecology. He wrote the report 'Freedom & Society in a Complex World' for the GEF. Affiliated with the University of Antwerp's Centre for Research on Environmental and Social Change, he is conducting research on the role of citizens' collectives in building a new economy

ABOUT THE AUTHORS

Natalie Bennett has been a Green Party member of the House of Lords in the UK since 2019, having been leader of the England and Wales party from 2012 to 2016. A feminist since the age of five (when she was told because she was a girl she couldn't have a bicycle), her agricultural science degree undertaken in her native Australia demonstrated the destructive nature of farming practice, without offering alternatives. Subsequently, while working on Asian studies and having done a Master's in mass communications, Natalie has been trying to build bridges between social science, the humanities and the physical sciences. As a journalist – more recently as an editor of *Guardian Weekly*, she has watched the descent into these last zombie days of neoliberalism.

Anne Chapman is former co-chair of Green House Think Tank. She worked as an environmental consultant, following a degree in biochemistry at Oxford University and a Master's in environmental science at Manchester University. She also did a Master's and then a PhD in environmental philosophy at Lancaster University. Her PhD, on the regulation of synthetic chemicals, was published in 2007 by Earthscan as *Democratizing Technology; Risk, Responsibility and the Regulation of Chemicals*. For Green House she has written about greens and science, housing, dealing with extreme weather events, climate jobs and community energy in the United Kingdom. She grew up in a rural part of South Shropshire and is now co-owner of a small area of partly bracken-covered hillside in South Cumbria.

Daniel Chavez is a Uruguayan anthropologist and political economist specializing in emancipatory politics and public policy. He is a senior researcher at the Transnational Institute (TNI), an Amsterdam-based think tank. He joined TNI in 2001 as coordinator of the Energy Project, looking at democratic and participatory alternatives to electricity privatization in the Global South. Before moving to Europe he had worked for almost a decade for the Uruguayan Federation of Mutual Aid Housing Cooperatives (FUCVAM). Daniel currently coordinates the TNI New Politics Lab. He has authored and edited a number of books, including most recently *Public Water and Covid-19: Dark Clouds and Silver Linings*, with Susan Spronk and David McDonald; *The New Latin American Left: Utopia Reborn*, with Patrick Barrett and Cesar Rodriguez Garavito; and *The Left in the City: Participatory Local Governments in Latin America*, with Benjamin Goldfrank. He holds a PhD in development studies from the International Institute of Social Studies of Erasmus University Rotterdam (ISS, The Hague).

Sean Currie was born the son of an oil rig worker in rural scotland and is now a 24-year-old social movement researcher, organizer and trainer. Over the last 8 years he has been active in diverse forms of activism, from Green party politics to community organizing, working to connect the needs of communities to high-level politics. Based in Brussels, he is currently finishing a Master's at Sciences Po, Paris, in trying to understand the European Union.

Dragan Djunda is a PhD candidate in the department of sociology and social anthropology at the Central European University. His research deals with the political ecology of energy transition in Serbia, with an emphasis on investments in hydropower and their economic and ecological consequences. He is a research associate at Polekol and Pravo na vodu, and a member of Elektropionir – the cooperative for citizens' solar energy based in Serbia.

Rand El Zein, born in 1991, is a researcher and climate activist from Beirut, Lebanon. She received her Doctorate degree from Universität Salzburg in Austria, where her thesis was titled 'Between violence, vulnerability, resilience and resistance: Arab television news on the experiences of Syrian women during the Syrian conflict'. Her research focuses on Arab media, cultural studies and eco-feminist practices in neoliberal societies and in the Arab Muslim world.

Lyda Fernanda Forero is a policy advisor at the Trade Union Confederation of the Americas (CSA-TUCA). She is an economist who carries out analysis and campaigning on topics including trade and investment policies, the architecture of impunity created for transnational corporations, and new trends in the financialization and commodification of nature and of life. Lyda is Colombian, and she holds a BA in economics and an MA in history at the National University of Colombia, where she was a teacher and a researcher. She was also a researcher for the Transnational Institute's Economic Justice, Corporate Power and Alternatives Programme.

Aleksandar Gjorgjievski is the president of North Macedonia's Association for Sustainable Social and Economic Development (Sunrise). Currently finishing a Master's degree in economy and finance, he has worked on diverse environmental topics such as waste management, air pollution, sustainable food systems and energy transition. For Sunrise he has written publications about decreasing air pollution and the geothermal energy perspectives for North Macedonia, and has authored articles on food, mobility, energy transition and pollution. Aleksandar is currently working with Sunrise to establish energy cooperatives in the region and is closely following the energy transition in Southeastern Europe.

Raúl Gómez studied Spanish philology but his working relationship with environmental issues began at Greenpeace Spain. He joined as a volunteer in 2002 and one year later joined the staff in the administration department. He also worked on the Forest and Coastal Areas campaign, staying on until 2010 by which time he was a management assistant. He founded the green party EQUO and the foundation of the same name in 2010/11. In 2016 he became the director of the foundation, which in 2018 changed its name to Transición Verde (Green Transition). He has been a member of the General Assembly of the Green European Foundation since 2016. He also directs a collection of books on the pioneers of environmentalism: Hojas en la Hierba (Leaves in the Grass).

Vedran Horvat (sociologist, journalist and activist) has been head of the Zagreb-based Institute for Political Ecology since 2015 and is a member of the board of directors of the Brussels-based Green European Foundation. From 2005 to 2015 he was director at the Zagreb office of the Heinrich Böll Stiftung, the German Green Political Foundation. Prior to that he worked as a journalist with the daily political newspaper Vjesnik and as a contributor to international newspapers and Transitions Online, based in Prague. He is author and editor of dozens of publications and articles on ecological sustainability, democracy and migration.

Anya Namaqua Links is an indigenous Namibian translator, researcher and writer. She mostly writes in her native language, Afrikaans, and in the official language of Namibia, English. Her areas of interest are democracy, development and the environment, the intersectionality of gender, tribalism and racism in contemporary African politics, and the participation and representation of minorities in post-colonial southern Africa. She holds a Bachelor's degree in political science and lives in Windhoek, the capital of Namibia, where she makes handmade soap and traditional food, and helps care for her extended family and stray dogs.

Robert Magowan is a policy advisor in government, activist and a core group member of Green House Think Tank. He was formerly the policy development co-coordinator for the Green Party of England and Wales, and together with the PCS trade union, he co-authored the report 'A Green New Deal for Gatwick', which follows the impact of Covid-19 on airport towns. He studied politics at Newcastle University and economics at Leiden University. He grew up in Armagh, Northern Ireland.

Sara Matthieu has been a member of the European Parliament (MEP) for the Flemish Green party Groen since 2020. She has been active in Green politics for over 21 years. She has held multiple positions within Groen, on the youth, local, national and European level. She has worked for Natuurpunt, the biggest nature conservation organization in Flanders. She is the former chief of the general politics team in the cabinet of Brussels Mobility Minister Bruno De Lille. Sara has built expertise in such fields as environment, mobility, energy and social politics and trade. She is an active member of the Belgian women's movement. She is a city councillor in Gent, where Greens are part of the majority. Sara has been active in European politics since 2002, first with the Federation of Young European Greens and since 2007 with the European Green Party. She has founded the European network of Groen together with the MEP Bart Staes. Since 2012, she has been a member of the European Greens committee.

Joaquín Nieto has been the director of the Spanish office of the International Labour Organization since 2011. He was previously the Confederal Secretary for Environment and Occupational Health of the Spanish trade union Comisiones Obreras (1991–2008), and he is the president and co-founder of the Trade Union Institute for Labour, Environment and Health (ISTAS). At the international level he has been a member of the European Union's Environment Advisory Forum (1995–2002); he co-founded Sustainlabour (the International Labour Foundation for Sustainable Development), of which he was president between 2004 and 2008 and honorary president between 2008 and 2011; and he has participated as a workers' representative in the United Nations Commission on Sustainable Development (1996–2007), in the Climate Summits (1995–2009) and in the United Nations Environment Programme (2005–2007).

Simo Raittila is the coordinator of the green think tank Visio. He has a Master's degree in social and public policy from the University of Helsinki and is currently researching the mental health risks of working-age people for his PhD in sociology. Prior to this he worked at Kela, the Social Insurance Institution of Finland, studying last-resort social assistance.

Adrián Tóth is a biologist by training, and he now works in the EU climate and environment sector. He is currently a project coordinator at the Green European Foundation. Adrián is based in Brussels where he has worked at the European Commission, in the EU office of the Covenant of Mayors, and for the Friends of Europe think tank. He is also co-founder of the grassroots initiatives Plastic Free Plux and Sustainable Brussels Walking Tours, both of which focus on promoting transformative cities and communities. He has an MSc in tropical biodiversity and ecosystems, with field experience in Europe and Southeast Asia.

Elina Volodchenko obtained a Master's degree in sociology at the University of Ghent, majoring in conflict and development in a global world. Her areas of interest include topics such as inequality, globalization, ecology, postcolonialism and intersectionality. Additionally, she has started a Master's degree in journalism at the Free University of Brussels.